W9-CKI-144

The Case of the Green Turtle

The Case of the Green Turtle

*An Uncensored History of
a Conservation Icon*

ALISON RIESER

The Johns Hopkins University Press

Baltimore

© 2012 The Johns Hopkins University Press
All rights reserved. Published 2012
Printed in the United States of America on acid-free paper
9 8 7 6 5 4 3 2 1

The Johns Hopkins University Press
2715 North Charles Street
Baltimore, Maryland 21218-4363
www.press.jhu.edu

Library of Congress Cataloging-in-Publication Data
Rieser, Alison.
   The case of the green turtle : an uncensored history of a conservation icon / Alison
Rieser.
       p. cm.
   Includes bibliographical references and index.
   ISBN 978-1-4214-0579-7 (hdbk. : alk. paper) — ISBN 978-1-4214-0619-0
(electronic) — ISBN 1-4214-0579-2 (hdbk. : alk. paper) — ISBN 1-4214-0619-5
(electronic)
   1. Sea turtles—Conservation—History. 2. Carr, Archie Fairly, 1909–1987. I. Title.
   QL666.C536R555 2012
   597.92'8—dc23   2011047301

A catalog record for this book is available from the British Library.

Special discounts are available for bulk purchases of this book. For more information,
please contact Special Sales at 410-516-6936 or specialsales@press.jhu.edu.

The Johns Hopkins University Press uses environmentally friendly book materials,
including recycled text paper that is composed of at least 30 percent post-consumer
waste, whenever possible.

For Jenna

If the world goes on the way it is going it will one day be a world without reptiles. Some people will accept this calmly, but I mistrust the prospect. Reptiles are part of the old wilderness of earth, the environment in which man got the nerves and hormones that make him human. If we let the reptile go it is a sign we are ready to let all wilderness go. When that happens we shall no longer be exactly human.

Bore through to the core of what is required and you see that it is an aggressive stewardship of relics, of samples of the original order, of objects and organizations of cosmic craft. This work will take stanch people, and the reptile can be the shibboleth by which they pass.

Archie Carr, *The Reptiles* (1963)

Science does have a major role to play in the future of conservation. . . . Nevertheless, science will need careful guidance and supervision from other disciplines; and even given the best of circumstances, the outcome of its efforts will remain for some time beyond the reach of scientific prediction.

David Ehrenfeld, *Biological Conservation* (1970)

Contents

Preface

I've long been intrigued by the social behavior of scientists who study marine organisms and the oceans. This fascination probably dates from my years as a "summer kid" in Woods Hole, Massachusetts, where one of the highest densities of research scientists aggregates every summer, and where a child could stand in line at the ice cream counter and see the knobby knees of a Nobel laureate. But in my professional work as a teacher of ocean law and policy, I noticed a curious contradiction. Despite their deep concern for the survival of the species and ecosystems they study, some marine scientists eschew involvement in environmental policy debates. A few even criticize their colleagues who step outside the ivory tower to use their research to protect nature or to tell the rest of us that we must stop our abuses of the biosphere. I wondered why. Why do some scientists urge their students to stifle their personal convictions and adhere to the model of the dispassionate and objective scientist? What are the risks to the careers of researchers who appear to "cross the line" into environmental politics? Does the elusive distinction between fact and value become further blurred? Do colleagues dismiss you as a grandstander?

These questions led me to the fascinating world of sea turtle biology and the divisive question whether mariculture or other commercial uses of sea turtles could be a hedge against extinction. I decided to write the history of this controversy for the guidance it may provide the current generation of naturalists, especially those who regard themselves as conservation biologists, and to preserve the story for future generations of conservationists from all disciplines. Biology, it seems, endows its practitioners with powerful insights and an almost irresistible urge to apply these insights to improve the world. But biology is also a contest of ideas and in today's world, sadly, often a dispiriting competition for recognition and funding. I hope this story will give all who work to save nature

a renewed sense of purpose by connecting their efforts to their coura-
geous and deeply human forerunners in the imperfect but essential world
of nature conservation. To my friends whose interests are in environmental law, I hope it reminds them that underneath every lawsuit lies a
dramatic clash of human hopes and values.

Acknowledgments

My debts in writing this book are many, particularly to the three turtle scientists who agreed to read my manuscript to make sure I made no egregious mistakes about turtle biology or ecology: Sally Murphy, David Owens, and Peter Pritchard. Holly Doremus, an expert on the role of science in the Endangered Species Act, took precious time from her sabbatical in Hawaii to read the text and make important suggestions. Notwithstanding their efforts, any errors that remain of either science or law are entirely my responsibility.

Others who shared their stories of turtles and turtle people, and helped me to find many others, include Brian Bowen, William Denevan, Kenneth Dodd, Marydele Donnelly, Karen Eckert, David Ehrenfeld, Judith Fitzpatrick, Jennifer Homcy, Julia Horrocks, George Hughes, Wayne King, Nancy Lewis, Colin Limpus, Frank Lund, Helene Marsh, Anne Meylan, Nicholas Mrosovsky, Ransom Myers, Pamela Plotkin, Henk Reichart, Carl Safina, Robert Schmidt, Jeff Seminoff, and Wayne Witzell.

Karen Bjorndal and Alan Bolten were gracious hosts at the Archie Carr Center for Sea Turtle Research, at the University of Florida in Gainesville. They pointed me toward the rich resources in the Carr and Harrisson Papers housed in the Special Collections at the Smathers Libraries, where numerous librarians and archivists made my first stab at archival research a pleasure. I'm particularly indebted to Special Collections librarian Florence Turcotte. Peter Eliazar of the Carr Center was also an enormous help, as was the Center's online sea turtle bibliography. I also received aid from the wonderful librarians at the Darling Marine Center Library of the University of Maine, Hamilton Library of the University of Hawaii, Cornell University Library, Bishop Museum Library, Minnesota Historical Society, Louisiana State University Library, the National Archive of Australia, the Sarawak Museum, and the State Libraries of Florida and Queensland, Australia. Material important to my story was graciously provided by Corey Malcom at the Mel Fisher Maritime Museum in Key West, Joan Langley of the Wright Langley Archives, Jim Lowry of the Australian Museum, Mishka Chisolm of the Cayman Islands National Archives, Roger C. Smith of Florida's Bureau of Archaeological Research, and Christine Horn of Swinburne University of Technology, Melbourne, Australia.

Judith Heinmann helped me decipher Tom Harrisson's handwriting and kindly shared recollections of her conversations with John R. Hendrickson about his work in Sarawak. Karla Kishinami and Mrs. Lupe Hendrickson also helped me understand Hendrickson's early research years. Regina Luna Rudrud shared memories of the later years of Cayman Turtle Farm and made sure I had the latest science papers on the Hawaiian green turtle. Sam and Peggy Fosdick offered their recollections of the people who worked at Mariculture, Ltd., and of Heinz and Judith Mittag. Their meticulously researched and engrossing book on Mariculture, Ltd., and Cayman Turtle Farm introduced me to the story and was an invaluable resource.

For legal and archival research assistance I thank Professor Alyson Flournoy, Kevin Sharbough, and Laura Chen Allen. For permission to use personal photographs, photographs of family members, and maps prepared or commissioned by their loved ones, I thank Robert Bustard, David Carr and the Carr Family Trust, Judith Fitzpatrick, Leslie Hendrickson, Karla Kishinami, David J. Parsons, Peter C. H. Pritchard, and Jack Rudloe. A fellowship from the Pew Fellows Program in Marine Conservation supported early research on this book, including a trip to a leatherback nesting beach in Costa Rica. Summer research grants and a sabbatical leave from the University of Maine School of Law were instrumental, as was the Dai Ho Chun chair from the University of Hawaii at Manoa's College of Social Sciences from 2006 to 2008.

Finally, I thank my own loved ones, Jenna and Les Watling, and my mother, Catherine Rieser Williams, for their support and forbearance as I told them my sea turtle stories.

The Case of the Green Turtle

From Seafood to Icon

In the field of marine conservation, one group of species garners disproportionate attention from researchers and government agencies. This group is the sea turtles. To be sure, sea turtles are fascinating creatures to study and observe in the wild. But their power to fascinate is not the only reason they get so much attention. As one observer predicted, by 1980 they had displaced the whales in the "Save the whales" rallying cry of the environmental movement. Sea turtles are now iconic species, emblems for the campaign to save the oceans.[1]

The sea turtles' iconic status, and the public stewardship that follows from it, are both premised on the idea that sea turtles are endangered, that some or all the remaining species are at some measurable risk of extinction. Yet a number of scientists active in sea turtle research today question whether most sea turtles are endangered. Most agree that some regional populations require human intervention to prevent their extirpation. And some have argued that we should restore other populations to their former role as keystone species in tropical and subtropical marine ecosystems. One particularly active group of turtle scientists argues that sea turtles are not facing even a medium-term risk of extinction. Nevertheless, all sea turtle species remain listed under the US Endangered Species Act of 1973, and all are classified as threatened under CITES, the international treaty controlling trade in wild species of plants and animals.[2]

The green turtle, *Chelonia mydas*, is the most heavily exploited of all of the marine turtle species and the first to achieve a degree of recovery because of conservation policies adopted more than thirty years ago. Currently, several conservation scientists are marshaling evidence that the green turtle is no longer endangered. Aware that this classification is both a scientifically derived status and a social construction, these scientists have a variety of motives and tactics. Some seek to demonstrate that

1

conservation interventions can work and that species can be returned to a nonimperiled state. Others believe that the total preservation strategy adopted in the late 1970s worked an injustice in some human societies; they seek to restore the green turtle to the status of an exploitable resource. Still others may be inspired by a desire to reprise and prevail in a philosophical debate begun by the first generation of sea turtle conservationists.

Under US law, only two populations of green turtles are considered "endangered"—the population that nests on the Pacific coast of Mexico and the one that nests on the east coast of Florida. The remaining populations, including the green turtle population of the Hawaiian archipelago, are considered "threatened." The International Union for Conservation of Nature (IUCN) Red List is the international classification scheme for species, and it has classified the green turtle as globally endangered since 1968. The specialists responsible for this classification argue within their ranks (and in their publications) whether the green turtle is really endangered globally.[3] But under the cover of this debate, they are actually reprising another debate that raged among turtle scientists during the 1970s: should the green turtle be commercially exploited or protected from all human consumptive uses until its role in tropical marine ecosystems is restored?

This book is a history of how this one globally distributed species of sea turtle—the green turtle, *Chelonia mydas*—came to be classified legally as an endangered species and why it remains so today. It is a history of the first generation of scientists who, fearing the green turtle would become extinct in their lifetimes, used the laws and treaties they helped devise in the 1970s to prevent that from happening.

A Surprising Conclusion

Steven L. Yaffee's 1982 book *Prohibitive Policy: Implementing the Federal Endangered Species Act* got me thinking about green turtles and the law. A student of natural resource policymaking, Yaffee was interested in learning how bureaucracies adopt one of the most extreme forms of government intervention—the absolute prohibition—a policy that bans certain human behavior in order to achieve a particular social goal. Prohibitory policy was increasingly the norm in the 1970s, but economists had begun to criticize its inefficiency, arguing that market-based incentives and taxes would be more effective strategies. In the face of this criticism, Yaffee was curious why, when, and how administrative bureaucracies adopt prohibitory mandates and use their scientific and technical expertise to legitimize them.

Yaffee's book focused on the prohibitory regulations that resulted when six species of plants and animals were classified as either endangered or threatened with extinction. These rules had been adopted in the first decade after the US Congress enacted the Endangered Species Act in 1973. The six species included two sea turtles; the other four were the sandhill crane, the Furbish lousewort, the Houston toad, and the now infamous snail darter. Of the six, only the green turtle had commercial value. How had government agencies handled the difficult job Congress gave them of prohibiting human activities in order to protect a wild species?

Not well, Yaffee concluded. And the turtle listing decisions were particularly revealing. The agencies had clearly considered economic interests even though the Endangered Species Act directs them to base their decisions solely on science. To avoid the more prohibitory policy that comes with the label of "endangered," the agencies had classified the heavily exploited and highly depleted green turtle as "threatened with extinction" rather than endangered. This classification gave them the flexibility to postpone adopting prohibitive policies for the politically potent commercial fishing industry.

This finding was not surprising. In my years of teaching environmental law, I'd seen dozens of court cases challenging agencies for going too far in seeking an elusive middle ground between protecting the environment and allowing industries to put useless things into it or take valuable things out of it. What struck me instead was Yaffee's concluding observation on the sea turtle listings. To Yaffee they were examples of a more unexpected decisional problem—one where "personal philosophy heavily influences scientific judgment, leading to conflicting positions on technical decisions."[4]

Could this be true? Why would decision makers need to fall back on personal philosophy to resolve technical issues about a species in peril? Was there any doubt that the green and other species of sea turtle were in danger of extinction? The Endangered Species Act requires listing decisions to be based on the "best available scientific data."[5] Surely the listing petition contained sufficient historical evidence to demonstrate that overexploitation and habitat destruction had caused the loss of one green turtle nesting population after another. Or was it that personal philosophies had somehow influenced the view of agency scientists and their advisors on whether a prohibitory policy was necessary to prevent the sea turtles' extinction?

The answers lie in a 1978 decision by the two US fish and wildlife agencies to list the green turtle as threatened (nearly endangered) and to prohibit import into the United States of farm-raised turtle products for use in soup and cosmetics manufacturing. This decision committed the United States to a species conservation policy based on preservation of individual species and their recovery to levels at which they would no longer be at risk of extinction. It closed off domestication as a conservation strategy, ending any further consideration of legal commercial use. With respect to the green turtle, there would be no further debate on the issue that is frequently aired today: whether a "sustainable-use" strategy might be fairer to human communities located near the habitat of the endangered species and, by increasing the value of the species to those communities, promote a broader ecological approach.[6]

In the 1970s, the idea of domestication appealed to a thoughtful minority of those scientists who at the time were studying sea turtle biology and ecology. To this group, turtle mariculture had promise to supply farm-raised turtle meat and shell to both local and distant markets and take pressure off wild green turtle populations, many of which were being exploited at unsustainable rates.[7] But the majority of sea turtle specialists opposed turtle farming; among them was Archie Carr, the acknowledged founder of sea turtle science and conservation.[8] Carr, however, grew tired of the domestication issue and became increasingly reluctant to engage directly in policy debates with colleagues or testify in person in legislative proceedings. Others in his circle of former students and close associates took on the task, often in a much more combative manner than Carr would have used. The argument took on a personal tone when the research at Cayman Turtle Farm, the only commercial farm in operation, was criticized, and agency proceedings found the farm's research contributed little to turtle conservation in the wild. The issue was never resolved on the merits because the farm was never granted an exemption from the Endangered Species Act. The effects of the dispute have lingered on among turtle specialists.[9]

Becoming Endangered

The green turtle is a globally distributed megaherbivore with rookery beaches scattered throughout the tropics and subtropics. The largest rookery beaches are along the Caribbean coast in Central America and on tiny islands on Australia's Great Barrier Reef. Significant rookeries also occur on the South Atlantic's isolated Ascension Island and on the tiny Indian Ocean islands of Europa and Tromelin, near the French territory of Réunion. The green turtle population that nests on the southeast coast of Florida is beginning to recover and now constitutes the second largest nesting population in the Atlantic Ocean.[10]

Known for their extensive migrations between their nesting locations and the seagrass pastures where they forage off Nicaragua, Brazil, Kenya, and Queensland, green turtles are long-lived, reaching reproductive maturity only after several decades. After mating, female green turtles go ashore at their natal beaches and lay several clutches of eggs over a period of weeks. They then migrate back to foraging areas for two or more years before returning for another nesting season. After incubating in nests buried deep in the sand, green turtle hatchlings emerge in frenzies and cross the beach and intertidal zone, running a gauntlet of predators and physical hazards. Once they reach the open ocean, they spend a few years drifting passively in ocean currents, foraging on tiny marine life inhabiting the sargassum weed that collects at the oceanic convergence zones. Juvenile green turtles then take up residence in coastal lagoons far away from their beaches of origin and later somehow find their way back to these beaches when they are ready to breed.

The green turtle is classified as endangered on IUCN's Red List of endangered and threatened wild species of plants and animals, and has been since the late 1960s. The current classification is based on a finding by sea turtle specialists that the global population has declined over the last three generations by more than 60 percent and therefore faces a measurable risk of extinction. The causes of the decline are many and vary by region, but the principal reasons are the international demand for green turtle meat and cartilage used for soup, which peaked in the 1960s, and the local consumption of turtle eggs. International trade in green turtles has existed since the European discovery of abundant Caribbean populations in the sixteenth century. The green turtle is exceptionally easy to capture when on its nesting beaches, and for centuries a highly efficient fishery was prosecuted by the skillful turtle fishermen of the Cayman Islands. From the 1830s until the 1960s, the Cayman Islanders captured turtles on the Miskito Bank, off the coast of eastern Nicaragua, and delivered the giant green turtles live to busy canneries and markets in the United States and Europe.[11]

Scientists began research and management activities on turtle beaches in the 1940s, first on three tiny islands off the coast of northwestern Borneo and then, in the 1950s, at Tortuguero, on Costa Rica's Caribbean coast. By the 1970s, tagging studies and restoration projects were in place at every significant nesting beach around the world. Scientists compared notes and devised conservation strategies when they met under the auspices of IUCN's Marine Turtle Specialist Group, which Archie Carr had been invited to chair by the founder of the World Wildlife Fund, Peter Scott. Scott, son of the famous Antarctic explorer Sir Robert F. Scott, was a painter and noted waterfowl conservationist, who popularized natural history by hosting a series of television programs for the British Broadcasting Corporation. Scott was an advocate of captive breeding of endangered species and had an early success with the Hawaiian goose, known as the *nēnē*, through captive breeding at his estate in Gloucestershire, England.

Going beyond their traditional role as taxonomic classifiers, the sea turtle specialists began to classify marine turtle species on the basis of survival probabilities, using a scheme devised by Peter Scott. They found that by the middle of the twentieth century the green turtle was again subject to a high level of exploitation, occasioned by the development in the 1960s of worldwide trade in calipee, the cartilage that joins the lower shell, or plastron, to the skeleton and was the principal ingredient of the hugely popular green turtle soup. The Nicaraguan government had abrogated its treaty with Britain and ejected the Caymanian schooners from their turtling grounds on the Miskito Bank. But the turtles were now being taken in large numbers from these same grounds by the coastal Miskito villagers. They were paid cash to deliver their traditional subsistence fishery to three new slaughterhouses that processed hundreds of turtles every day for export to the turtle soup manufacturers and leather goods companies of the world.[12]

When the US Fish and Wildlife Service published a list of the species of marine turtles it considered endangered in 1970, the agency did not consider the green turtle to be in enough peril to warrant listing alongside the leatherback, the Atlantic ridley (now known as Kemp's ridley), and the hawksbill turtles. This was despite the fact the Marine Turtle Specialist Group classified the green turtle as at risk of extinction. Three Central American countries harboring the largest green turtle populations were in a race to profit from the growing demand for green turtle products, constructing their own slaughterhouses and allowing their fishermen to ship turtles live to soup manufacturers in the United States. Many turtle dealers saved shipping costs by buying from fishermen who would slice open and remove the five pounds of calipee from the female turtles when they came ashore to nest. Carr's nongovernmental organization, the Caribbean Conservation Corporation, organized to support his research and conservation activities, had tried but failed to broker a tripartite treaty for a moratorium on turtle harvesting that would give the neighboring countries time to develop a regional management plan.

Down on the Farm

Perception of the green turtle's survival prospects changed dramatically in 1970. A small group of entrepreneurs seemed poised to satisfy the demand for green turtle products through mariculture—raising turtles as aquatic livestock. Having made independent fortunes in the manufacture of cigar boxes and the industrial-scale rearing of broiler chickens, the principal investors joined forces with a Florida marine biologist, Robert Schroeder, who had worked with Archie Carr and was inspired by his musings that farming might be necessary to save the green turtle. Robert Schroeder believed the green turtle was the buffalo of the sea.[13] If it were allowed to graze the seagrass meadows of the Caribbean coastline, people could tap the tremendous nutritional value of seagrass by eating green turtle meat. The turtles would be domesticated and raised for market so efficiently that they would undercut the market for wild turtles, especially if that market were made illegal under the new treaty, CITES, which would soon control international trade in endangered plants and animals.

Archie Carr had mentioned the idea of turtle farming in his 1963 general readership book, *The Reptiles*. He told the story of how a group of publishers and philanthropists had formed the Brotherhood of the Green Turtle after reading of its plight in an earlier book of Carr's, *The Windward Road*.[14] The Brotherhood funded Carr's Operation Green Turtle, an effort to restore extirpated rookeries throughout the Caribbean basin and establish a pilot turtle farm in the Bahamas, where green turtles could be pastured on seagrass beds like "aquatic cattle." Carr wrote, "If they are successful, the green turtle may become one of the first marine vertebrates to be successfully cultured for food." But he expressed ambivalence about farming as an indirect conservation strategy. He much preferred that humanity acknowledge the intrinsic value of reptiles and save them for that reason: "The only way is to name the real obligation clearly, to say without hedging that no price can be set for things that have to be preserved."[15] Species like the New Zealand lizardlike tuatara must be saved "so that people can sing them out of their holes. Only then are you ready for the harder jobs, like justifying a future for snakes, which have no legs, hear no music and badly clutter subdivisions. Bore through to the core of what is required and you see that it is an aggressive stewardship of relics, of samples of the original order, of objects and organizations of cosmic craft. This work will take stanch people, and the reptile can be the shibboleth by which they pass."[16]

Pursuing the vision of an aquatic buffalo range, Schroeder and the entrepreneurs built their green turtle farm on Grand Cayman Island, the former home of the world's most effective turtle fishing fleet and former site of the vast turtle populations that Columbus had remarked upon in the journals of his voyages. Building their breeding stock from wildcaught turtles from Costa Rica and eggs from Suriname, Costa Rica, and Ascension Island, the businessmen planned to restore wild green turtles by supplying the market with a superior product. To stop the farming idea from catching on, species survival activists who admired Archie Carr and shared his preservationist values petitioned the government to list the green turtle as endangered and prohibit trade in all green turtle products. Mariculture, Ltd., the company that built upon Robert Schroeder's idea, mounted a vigorous defense, insisting that captive-raised turtles were not endangered and that a commerce-based conservation strategy deserved to be tried. Mariculture went bankrupt, however, before it had a chance to prove Schroeder's "buffalo of the sea" theory.

While the listing was pending, a former research assistant of Carr's, Wayne King, who had filed the listing petition while serving as director of conservation for the New York Zoological Society, lobbied state legislatures in New York and California to ban the importation of turtle skins and oil, preventing Mariculture from selling its products to cosmetics and shoe manufacturers. Meanwhile, another turtle drama took place in the cabinet room of Florida's governor. A college student named Frank Lund, whom Archie Carr had encouraged to document green turtle nesting on Florida's east coast, convinced the governor to enact a forty-two-inch minimum size rule for green turtles. The governor did so over the objection of Florida's director of marine research and fisheries, effectively shutting down what remained of the Cayman Islands–Key West trade in turtles from the Miskito Bank.

Court of Last Resort

The centerpiece of the green turtle's history as an endangered species is the case of *Cayman Turtle Farm, Ltd., v. the Secretaries of Interior and Commerce*, decided by John H. Pratt, a judge on the federal district court in Washington, DC, on May 29, 1979. The plaintiffs were the new owners of the turtle mariculture operation on Grand Cayman Island. A physician from Germany and her industrialist husband (who had made a fortune selling her patented personal care product, the OB tampon, to the Johnson & Johnson health products giant), Judith and Heinz Mittag, had purchased the bankrupt farm because they believed the green turtle was at risk of extinction and they wanted to do something about it.

The Mittags did not contest the listing of the green turtle as threatened. They disputed only the Interior and Commerce secretaries' decision to prohibit importation and trade in farmed green turtles in the United States. Without transshipment through US ports, their plan of selling farmed turtle meat and shell to stanch the drain on wild green turtle populations and use the proceeds to fund research would be null and void. They had faith that the agencies administering the Endangered Species

Act would recognize the merit in their business plan and its consistency with the policies underlying that law. They asked the agencies to exempt farmed turtle products from a prohibitive mandate designed to protect wild species from extinction. After all, the sole objective of Cayman Turtle Farm was to use commerce to protect wild turtles from the unremitting exploitation they had suffered for centuries.

The secretaries had not had an easy time deciding to exclude farmed turtle products from the United States. Nor had they made the decision in a vacuum; some of the most eminent biologists in the world had offered their opinions, and had stated them in no uncertain terms. No scientific uncertainty led them to hedge or to qualify their recommendations. But these eminent views were on both sides of the issue. And the issue was fundamentally not one of biological science. The question was this: how would markets and people behave once the turtle farm made green turtle products widely available but charged luxury prices for them to recoup the costs of production? Would demand for green turtle soup again skyrocket, as it had in the 1960s? Would the price of tortoiseshell jewelry tempt people into poaching nesting turtles in newly established nature reserves? There was no "best available scientific information" on these questions, so people had to speculate and surmise rather than draw conclusions from empirical analysis or theoretical principles. And as Steven Yaffee concluded in *Prohibitive Policy*, they fell back on philosophical biases when there was no evidence to rely on one way or the other.

The judge who heard the case of Cayman Turtle Farm against the secretaries of interior and commerce upheld the agencies' decision to deny an exemption from the import ban for maricultured products. Judge Pratt of the federal district court of the District of Columbia relied largely on a principle of judicial review that requires judges to defer to the technical and scientific judgments of government agencies as long as those judgments are reasonable and based on the available scientific information and not on considerations that are outside the scope of the law they are charged with implementing. The judge deferred to the agencies' judgments that turtle mariculture would ultimately harm wild turtle populations. He did so because he could not see in the records of the jointly made decision what had really happened in the agencies' negotiations: the Interior Department staff had followed Archie Carr's lead in opposing commercial turtle farming on philosophical grounds, and the Commerce Department staff had struck a compromise that minimized restrictions on commercial shrimp fishermen who also catch sea turtles.

When Cayman Turtle Farm appealed Judge Pratt's decision to the court of appeals, the farm lost by a very slim margin. As it happened, the farm's lawyer had decided to appeal only a narrow issue of statutory interpretation. He did not challenge the agencies' technical conclusion that turtle farming would in fact lead to more killing of turtles in the wild and that the research on the farm did not benefit conservation. He probably should have. In a private note to the chief justice of the court of appeals, the appellate judge wrote: "The logic of this case is appalling. The agency contends that breeding large numbers of endangered species in captivity and thus greatly increasing the number of such species tends to further endanger the species."[17]

Conservation's First Social Scientists

Archie Carr opposed commercialization of green turtles in part because of what he had learned from the other scientists and writers whom he inspired with his books. These included the novelist and naturalist Peter Matthiessen, who wrote the haunting and dreamlike novel *Far Tortuga*, about the end of the Caymanian green turtle fishery and the loss of men and ships to storms on the Miskito Bank. Carr was also influenced immeasurably by two cultural geographers from the University of California at Berkeley. The first geographer was James J. Parsons, who wrote an influential book, *The Green Turtle and Man*, at the beginning of the 1960s detailing the exploitation and the global demand. The second geographer was Barney Nietschmann, who met Carr as a graduate student and got the idea for his dissertation from him. Nietschmann's study of the Miskito Indians, the turtle people of coastal Nicaragua, showed Carr that commercialization of the green turtle had tremendous social as well as biological costs.

Another person who influenced Carr was a self-trained polymath named Tom Harrisson who, as curator of the Sarawak Museum, administered an unusual turtle egg ordinance in the British colony on the island of Borneo. With the help of a young American zoologist, John R. Hendrickson, Harrisson pioneered the now-common techniques of tagging breeding adults, relocating eggs to artificial hatcheries, and protecting hatchlings from predation as they make their way beyond the surf zone. Before World War II, Harrisson co-founded the method of "mass observation" in sociology, documenting the everyday lives of Britons. But Harrisson cared little for the theoretical foundations of his survey methods. He was more interested in species conservation and archaeological discoveries in Southeast Asia than in making a name for himself in any particular subdiscipline of the social sciences.[18]

Tom Harrisson was co-chairman with Carr of IUCN's Marine Turtle Specialist Group. Because Harrisson held the same cultural values as the British members of IUCN such as Peter Scott, he was very effective in that role. Despite his many distractions, he marshaled the force of the new conservation institutions to the cause of sea turtle survival.[19] Harrisson drew also on his English upbringing and understanding of human behavior to campaign to depopularize green turtle soup in Britain and Germany. He then helped to devise guidelines for the commercial use of sea turtles but died suddenly before the Marine Turtle Specialist Group could reach consensus on the divisive debate over turtle farming.

A Shared History

The history of an endangered species has the potential to teach us several things. It can help us understand the social institutions and processes by which wild species are determined to be at risk of extinction. It can also show how science informs the concept of species extinction and how research scientists interact with political institutions that govern human-environmental interactions. In particular, such a history can identify the specific scientific facts that helped change the social construction of a species, from an animal that symbolized wealth and abundance to one that represented the plight of all imperiled life on earth. It allows us to consider how scientific facts and philosophical values interact in these social and political institutions. In this sense, an endangered species history is an environmental history, showing us how human society both alters its environment and seeks to rectify the alterations brought about by its own and previous generations.

The history of the green turtle as an endangered species turns out to be a history of a generation of scientists who sought divergent ways to prevent the green turtle from becoming rare or even extinct, using laws and treaties they helped to write and their abilities to appeal to the public's growing curiosity and admiration for the animal kingdom. In a sense it is a group biography of the people whose lives intersected through concern for the survival of the green turtle. It is the story of how their different philosophical values threatened to tear apart the scientific brotherhood of the green turtle and yet transformed the green turtle's social value for decades and perhaps for all time.

CHAPTER ONE

Turtle Kraals and Canneries

Silas Stearns was the first federal fisheries agent to venture into the state of Texas to study the fishing industry. Stearns worked for the US Commission of Fish and Fisheries, a body created by the US Congress in 1871 to determine the causes of the scarcity of food fishes along the Atlantic seaboard. As the scarcity of fish spread to other coasts, the fish commission's mission grew to include investigations of the Pacific, Great Lakes, and Gulf of Mexico fisheries. The fish agent's job was to accompany the US census takers, canvas the fishery resources of the state, and make a report to the US fish commissioner. The commissioner would then decide "what protective, prohibitory or precautionary measures were required" to restore them.[1] The measures might include hatcheries and stocking and introduction of wild fish from other waters.

Stearns's report told of finding great potential but little activity. "Nowhere do the rich Southern fauna find a more genial habitat," he wrote in 1880 to commissioner Spencer F. Baird, "and in few localities could man levy upon the sea a heavier tribute of delicious fish and mollusks to supply his table." Why, he wondered, were residents of the Gulf Coast catching so little? Fewer than 300 fishermen worked full-time in waters teeming with sea trout, mullet, croakers, crabs, and oysters. Up north, the fishermen of only one state, Massachusetts, had landings that were five times more valuable than those of all the Gulf states combined.[2] And Texas fishermen lagged far behind their counterparts in the other Gulf states, especially Florida.

The Texan fishermen did not take long to catch up. By 1890, almost 1,400 commercial fishermen fished from shore and plied the shallow bays inside the barrier islands in schooners, sloops, and catboats, catching almost fifty different forms of sea life for the growing market. Farmers and cattle hands were switching to fishing, joined by immigrants from Europe and Mexico, to gather the abundant sea life. One of their quarries was the green turtle (*Chelonia mydas*). The beef packinghouses that lined the shores of Aransas Bay had started canning a few green turtles in 1869. Ten years later, Mr. C. M. Hadden opened a turtle canning factory in Fulton, on Aransas Bay, buying turtles from local fishermen.[3] When the business proved good, he replaced the cannery two years later with a much larger factory. For the next fifteen years, the Fulton cannery processed about 1,000 turtles a year into 40,000 two-pound cans of turtle meat, and 800 two- and three-pound cans of green turtle soup.[4]

By 1890, catches of green turtles had increased twenty times over the paltry 24,000 pounds landed in 1880, and much of it was landed at the cannery in Fulton. The green turtles not processed into soup in Fulton, Texas, were sent to Galveston to be shipped live by steamer to the Fulton Fish Market in New York City or sold to the market in Corpus Christi. Fishing ports along the Texas coast near Corpus Christi soon all had "kraals," or turtle enclosures, built beneath the piers. These 500-square-foot enclosures were built with rough poles driven into the seafloor and connected at the top by a timber. The workers fed fish scraps and algae to the captive turtles until they could be shipped to market. They would be lifted by block and tackle attached to a swinging arm, put in shipping crates, and loaded onto side-wheel paddle steamers and schooners waiting along the pier for shipment to the canneries or to more distant markets.[5]

Soon government agents described Gulf fisheries as "remarkable for their recent wonderful growth," with plenty of room for further expansion. Texas fisheries were almost 70 percent more valuable than the average for the five states that bordered the Gulf of Mexico. The green turtle contributed a large part of that new wealth, ranking fifth among the dozens of species the fishermen were landing.

The green turtles appeared from the south in April and stayed until November. As the weather cooled, the turtles returned to the south for the winter, fattened from their months of grazing the seagrass beds of Aransas Bay, Matagorda Bay, Galveston Bay, and the lower Laguna Madre. Fishermen caught them by setting nets near the major passes within the bay, as they returned to the grass beds from their nightly resting spots in the deeper bay waters. Fishermen from Port Isabel in the lower Laguna Madre would post a lookout on the mast of the fishing scow and then set a net across the turtle's path.[6]

Charles H. Stevenson was the second fisheries agent to survey the coastal fisheries of Texas. On his visit in 1891 he tried to learn more about the catching and processing of green turtles for the soup canneries by talking to as many fishermen and dealers as he could find. He found it difficult to obtain catch and processing data from them that was reliable enough for a report to the US fish commissioner. So instead, he collected data on the number of turtles that were shipped from the region, estimating a total catch from Texas in 1890 at well over half a million pounds. Most green turtles were caught in Aransas Bay, but fishermen also brought them in from Matagorda Bay and the lower Laguna Madre. Stevenson paid a lengthy visit to the Fulton Canning Company and reported that it produced 40,000 two-pound cans of turtle meat in 1890 from the green turtles caught by fishermen of Mustang Island and environs. Stevenson found that Texas longhorn cattle were being taken in droves to packinghouses in Kansas, with the Texas packinghouses increasingly processing the green turtles that gathered to feed off wastes from their operations.[7]

Stevenson also learned that the green turtles processed in or shipped from Fulton, Texas, were not a local species. It was clear they were seasonal migrants foraging the dense seagrass beds of Aransas Bay. Green turtles were not nesting in Texas, and therefore there was no commercial turtle egg industry as there was in Florida.[8]

By the time Stevenson made his report in 1893, the Texas green turtle fishery was already in decline. In the town of Indianola, on Matagorda Bay, Harrison and Company, the purveyor of Harrison's Fresh Green Turtle, and Turtle Soup, had closed its packing plants in 1886. The business was shuttered along with the rest of the town, after the second hurricane in ten years struck the town. The storm demolished the canneries and the holding pool at Blind Bayou, where local children had watched 400-pound green turtles mill about before taking their final trip into the packing plant. Gone too was the dock where steamers of the Southern Steamship Company, owned by Charles Morgan of New York, loaded live turtles and cases of canned turtle meat and soup for the voyage from Indianola to Galveston and New Orleans.[9]

When the hurricane destroyed Harrison and Company in Indianola, commerce in green turtles moved to Corpus Christi. The largely foreign-born fishermen who had been catching green turtles with nets in Aransas Bay and Laguna Madre moved north and diversified into oysters, and the packing plants into chickens, game, oysters, and fish. Catches of green turtles in Texas were dropping sharply, so that by 1890, the turtle business in Texas was merely a third of the green turtle fishery in Florida's counties along the Gulf Coast. In 1896, the Fulton cannery that had made turtle meat an inexpensive staple of the local diet moved to Tampico, Mexico, to be closer to what remained of the green turtle population along the coast.

The winter of 1899 brought the second severe freeze in five years and dealt a death blow from which the Texas turtle fishery never recovered.[10] The freeze hit the turtle fishery before anyone had a chance to learn where the turtles came from, how many there were, and how many young they produced—information that would have been useful in managing the fishery. When the green turtle fishery failed to recover from the freeze, no one at the US Fish Commission or its counterpart in the state of Texas knew why. It might have been overexploitation, but then again it could have been pollution of the bays or prolonged cold temperatures or a combination of these factors. Because they knew little of the green turtle, they did not know of its long life, nor did they realize the fishery was depleting the large juveniles, a life stage the rapid loss of which would prove to be more damaging to the population than a similar depletion of eggs or small juveniles.[11]

Stevenson did not recommend any regulations to the US commissioner in his 1893 report even though he knew that fellow fish agents reviewing Florida's turtle fisheries had been alarmed enough to recommend immediate regulations. He was simply at a loss as to what to recommend, given what little was known of the origins and size of the green turtle population of the western Gulf of Mexico. The turtles in Texas were overexploited, but that was probably happening everywhere along the green turtle's migration route from Mexico to Texas, and perhaps beyond. Could the US Fish Commission really deal with that kind of fishery problem? Stevenson knew that the commission was more inclined to recommend culturing and restocking a depleted aquatic species than regulating a group of fishermen who were wary of federal authority. He wrote: "The work of fish-culture has met with so much success that many persons consider it more advisable to spend money for the hatching and distribution of fish, with a view to making them so abundant that they may be caught without risk of depleting the resources, than to spend an equal amount in policing and restricting the fisheries with a similar object in view."[12]

The head of the Texas Fish and Oyster Commission, Isaac P. Kibbe, had also seen the green turtle fishery collapse and in 1895 had pushed his legislators to enact a law. The Texas turtle law had three elements. First, it required all turtle fishermen to get a license so the state could learn how many there were. Second, the Texas law banned fishing by drag seine and set nets between April and October in waters "declared to be breeding grounds for fish, green turtle and terrapin." Kibbe defined the closed zone as extending all along the Texas Gulf Coast, from Turtle Bay in the northeast corner of Galveston Bay down to the Laguna Madre near the Mexico border. Third, the law prohibited landing turtles weighing less than twelve pounds. Commissioner Kibbe personally inspected the catches for violations of the twelve-pound minimum limit.[13] But Kibbe had few people to patrol the entire coastline, and the small catches did not seem to warrant aggressive enforcement. The six turtle canneries were almost all closed, and live shipments were few and far between. Fishermen were selling mostly for the local market.

Isaac Kibbe did not give up on the green turtle fishery, instead hoping it could be revived through active conservation. Kibbe disagreed with the federal officials who believed that overexploitation of the turtles in Texas waters had caused the decline. He believed the cause of the decline was the capture of nesting females and the harvesting of their eggs elsewhere—"in Mexico, Central America, the West Indies, and to some extent in Florida." Kibbe beseeched these governments to protect the turtles from overexploitation at the same time he fought for regulation to end turtle captures in Texas.[14]

But the US Fish Commission did not yet see the need to seek international cooperation to fulfill its mission of conserving America's food fisheries. The commission was not in the business of negotiating international agreements to conserve any fisheries, much less a fishery based on the ephemeral green turtle. Its mandate was to study America's fisheries and, where necessary, augment them through hatcheries. Spencer Baird, the first commissioner, had started fish culture investigations at the commission's first fisheries laboratory, in Woods Hole, Massachusetts. Was the green turtle a good candidate for cultivation? G. Brown Goode, Baird's long-time assistant and successor in office, said that it was probably easier to make more fish than to try to convince fishermen to take fewer.[15]

While Commissioner Kibbe was measuring green turtles in Texas, someone in Washington, DC, was worrying about sponges and oysters in Florida. In February 1895 the following resolution passed the US Senate:

Resolved, That the Commissioner of Fisheries is hereby directed
to make inquiry in reference to the extent, methods, and present
condition of the coast fisheries of Florida—more particularly the
sponge and oyster fisheries—and to report as to the desirability of
establishing a station for investigation, experiment, and fish-culture
at some suitable point on the coast.

The intent was clear. Congress wanted the US Fish Commission to find out what ailed Florida's commercial fisheries and whether the problems warranted construction of a federal research station somewhere in the state.

The fish commission already had quite a bit of data on the Florida fisheries. Most useful among this information was the 1885 report of the captain of the fish commission's steamer, *Albatross*, on a three-month survey of the waters around Florida. Captain J. W. Collins and his crew of biologists and fish agents called at several fishing ports, including Key West. Congress had previously directed the commission to make a "special investigation relative to the extermination of migratory fishes of the Indian River, Florida" in the appropriations bill for 1896; and the commission's report to Congress for 1895 had considered the suitability of Biscayne Bay as a site for a marine hatchery and experiment station.

In 1896 President Grover Cleveland appointed a new commissioner, John J. Brice, a retired Navy officer from California, who was the cousin of a senator from Ohio. There had been other contenders for the job, including the superintendent of the New York Aquarium, Tarleton H. Bean, who had worked for many years at the National Museum and had collaborated with G. Brown Goode in preparation of an exhaustive study of the US fisheries and fishing industry.[16] Supporters of other candidates for the job argued that Brice did not meet the requirement that the commissioner be "a person of proved scientific and practical acquaintance with the fishes of the coasts."[17] The Senate nevertheless confirmed his nomination, and Brice set out to make the required reports.

Brice appointed several of his best staff, including Dr. Hugh M. Smith, to return to Florida to follow up on the study made during the *Albatross* cruise. Smith and his team would assess whether conditions in each fishing region of the state had changed and describe for Congress the natural history of Florida's fisheries as well as their economic value. On this basis Brice would decide whether to recommend any measures to Congress.

Fishing in Florida was rapidly becoming a major industry, placing the state tenth among all the states in landed value, but it had a tremendous variety of fishes and other marine life, and the potential for growth was enormous. Florida was the only state with a sponge fishery, so it naturally led in that category. But it was also the leader in production of mullet, red snapper, pompano, kingfish, and green turtle.

To answer his critics, Brice authored the report himself and delivered it to the Senate in January 1897. He told Congress that Florida still had several thriving fisheries, but almost all were showing signs of distress. And when it came to the green turtle, it looked like Florida's fisheries were going the same way as Texas fisheries had.

Congress learned from Commissioner Brice that unlike Texas, Florida was blessed with a nesting population of green turtles, as well as abundant seagrass pastures that provided forage for juvenile migrants from other rookeries. When the market demand was high due to the American and European taste for turtle soup, Florida's fishermen readily captured both the adult turtles, when they gathered to breed near the Florida Keys, and the nonbreeding young adults foraging on the shallow grass flats off the Gulf Coast or in the Indian River Lagoon. Others raided the nests or turned nesting females over and cut the eggs out of their bodies. By 1895, the center of Florida's commercial fisheries was Key West, and its most prominent fishery was for the green turtle.

Turtle fishing and egg hunting had a long history in Florida. Historically, the largest green turtle rookeries were found in the remote keys and islands at the very end of the chain of islands that extended from the southern end of the Florida peninsula. It was here in the Dry Tortugas that the Florida turtle fishery had begun and the English palate was introduced to the taste of green turtle. An English explorer named John

Hawkins visited the Dry Tortugas in 1563 and wrote in his journals that the turtle he had consumed tasted "much like veal."[18] But it was the Spanish explorer Ponce de León who set the precedent for large-scale green turtle exploitation. Calling there in 1513 during the nesting season, his men captured 170 green turtles to serve as fresh meat for the voyage back to Europe. In exchange, he named the area the Tortugas, adding the descriptor "Dry" to warn his fellow voyagers that the islands lacked fresh water.

In the 1600s and early 1700s, green turtles were captured throughout the West Indies and loaded on Spanish exploring vessels to provision the trip homeward. By 1620, the English colony at Bermuda had fished out the local population of green turtles, and the colonists' enactment of a minimum size ordinance had failed to stem the loss.[19] Bermudian fishermen had to range farther afield to find turtles to sell to passing British and other vessels. They found very large nesting populations in the Florida Keys and along the east coast of Florida. But they were in competition with fishing parties sent ashore by ships sailing the passage between Florida and Cuba on their way to and from the Gulf of Mexico.

English schooners from the colony of the Bahamas soon joined the hunt and exploited heavily the green turtle populations that grazed on the seagrass beds off the east coast of Florida, off Cuba, and at the Dry Tortugas. As early as the 1730s, they crossed the Florida Strait to cut timber, salvage wrecks, and catch green turtles.[20] By the 1750s, the English court's favorite delicacy was green turtle soup made from turtles taken from the beaches and grass flats of Florida, Cuba, and Jamaica.[21] Once the rookeries at the Dry Tortugas were gone, the Florida fishery depended entirely on migrants from distant shores.[22]

Shortly after Florida became a US territory in 1821, the territorial governor sought relief from the Bahamian wrecking and turtling industry. He wrote John Quincy Adams, who, as secretary of state, was responsible for foreign affairs, complaining that Bahamians were taking very large quantities of "the finest turtles." He asked Adams to somehow create a monopoly for Florida's wrecking, turtling, and fishing industries.[23] Rather than negotiate a treaty that would most likely entail concessions, Adams got Congress to pass a law requiring all cargo and vessels salvaged in US waters to be taken to the nearest port of entry.

Salvage taken from the Florida Keys would need to be judged by the newly established salvage court in the tiny settlement at Key West. Shortly thereafter, the cutters of the US Revenue Service were harassing Bahamian fishing and turtling schooners and ejecting them from US waters. When the Bahamas governor tried to negotiate treaty fishing rights for its vessels, he was rebuked, prompting many of his best wreckers and turtlers to move lock, stock, and barrel to Key West.

The green turtle fishery thus became an American fishery and, centered at Key West, the most important commercial fishery in the young state of Florida. The American turtlers used the same fishing methods they did when they were Bahamians. They set large-meshed tangle nets 300–600 feet long and 16–24 feet deep across the channels among the keys or floated the nets above the turtles' resting holes along the edge of the reefs.[24] The turtles were entangled when they surfaced to breathe. The fishermen removed them the next day, delivering them to the kraals they had constructed beneath the wharves where schooners in the coastwise trade would load and unload their goods. Fishermen also captured green turtles during their summer nesting season by turning them on their backs and by pegging them in the water. This method involved harpooning the turtles through the carapace with a detachable bark that could be retrieved along with the turtle. Turtles captured in this way were not shipped north but were eaten instead by local people.

The Key West fishery was a year-round fishery extending throughout the Keys and Florida Bay, taking both juvenile and adult turtles.[25] The very small turtles, ranging from six to sixteen pounds, and the very large ones of more than two hundred pounds sold for between six and eight cents for local consumption.[26] The rest were sent north by steamer to the fish markets in New York, Boston, and European cities.

By 1880, ten turtling schooners employing roughly sixty men sailed out of Key West. This number dwindled to five or six large vessels by 1885, employing half as many men, with catch rates ranging from several dozen in only a few days, to zero in the course of a month.[27] But green turtle remained in the top three fisheries in 1895, with total landings of 337,000 pounds valued at almost $17,000. Egg collecting garnered an additional unreported amount.[28]

The green turtle fisheries in the other parts of Florida had a different history. Fishing for green turtles in the Indian River Lagoon got started

FIGURE I Dock workers loading green turtles for shipment from Florida to Fulton Fish Market, New York City, 1898.
Florida Photographic Collection, State Library and Archives of Florida, image number PR12653.

when a vessel from Connecticut looking for fish to sell in Savannah, Georgia, ventured into Florida's eastern coastal waters in 1878. Soon joined by local vessels, the new fishery caught an estimated 1,600 turtles with only eight men. By 1880, twice that number of fishermen was employed to catch 1,400 turtles.

In 1886, a new railroad line from Jacksonville made it easier and cheaper to transport the green turtles north, so the smaller catches were still profitable. One turtler, Mr. Charles Pearke, reported catching 2,500 turtles using set nets. By 1895, his catch had dropped to only 60 turtles at an average weight of only thirty-six pounds, down from the previous average of fifty to sixty pounds.[29] The fishery was catching younger juveniles as the older year-classes were fished out.

When an unusually severe winter in 1894 stunned or killed hundreds of juvenile green turtles in the Indian River Lagoon, the local citizenry collected every single turtle. This one bountiful harvest signaled the end to the Indian River Lagoon green turtle fishery. Fishermen also discovered a seasonal population of green turtles in the Cedar Keys on Florida's west-central coast on the Gulf of Mexico. After all the larger turtles were fished out, the fishery targeted the juvenile migrants who arrived in spring to feed on the banks of seagrass that surrounded the keys.

By the time New York chef Armand Granday moved to Key West in 1890, the green turtles landed at the wharves were all from distant waters, off Mexico's Yucatan Peninsula and other parts of the Caribbean.[30] Granday knew how popular clear green turtle soup had become among America's wealthy classes, having been the private chef to a New York financier. Banking on the idea that turtle soup would appeal to Americans who dreamed of moving up the social ladder, Granday opened a soup cannery on the wharf at the foot of Margaret Street in Key West. The labels on his cans implied that the turtles were a product of the waters around Key West, but within less than ten years, every green turtle that Chef Granday's factory made into soup was either from the Miskito Bank off Nicaragua—the product of the colorful Cayman Island turtle schooners that were known as "turtle boats"—or was caught on the beaches of Costa Rica.[31] When Norberg Thompson bought the Granday cannery in 1910, he had the master shipbuilders on Grand Cayman Island build several fast new schooners for the expert turtle captains to sail. The largest one was named after his general manager, A. Maitland Adams.[32] The record shipment occurred in 1919, when three turtle schooners brought in 750 green turtles from the Miskito Bank.[33]

William C. Schroeder, the young agent of the Fish Commission assigned to Key West, described how the giant greens of the Caribbean were converted into turtle soup. Swimmers entered the kraals, which stored up to 500 or more large green turtles. "A rope noose is dangled in the water until it catches the flipper of a turtle as it comes up to breathe. The rope

|T is a well-known fact, substantiated by such eminent culinary authorities as Urbain Dubois and Chas. Ranhofer, that green turtles, after living out of water for any length of time, lose part of their delicious flavor and acquire a fishy taste. *The quicker the turtle is cooked after leaving its element, the more delightful the flavor.* For this reason, we put up our soups at Key West, where the green turtle is at its best and thrives on the wonderful supply of food that Nature has there so abundantly provided for its benefit. Caught in the neighborhood, the turtles are taken from the sea directly to our kettles, and to this fact is due the well-known superiority of our soups. What champagne is to other wines, green turtle is to other meat. Prepared by our special process, nothing is a greater delicacy than green turtle soup made from selected young turtles cooked where caught. The difference in the flavor of our soup and those of competitors, who prepare their product hundreds of miles from where the turtles are caught, is well-known to epicures and connoisseurs.

NET WEIGHT 10 OUNCES

A. GRANDAY'S FINE GREEN TURTLE. KEY WEST, FLA.
DIRECT FROM SEA TO THE KETTLE.

TRADE MARK

CLEAR GREEN TURTLE SOUP

FIGURE 2 Granday's turtle soup label used by Thompson cannery, Key West, Florida.
Mel Fisher Maritime Heritage Society.

is at once hauled taut and the animal is hoisted to the dock, the precaution being taken to render it helpless by turning it on its back before the noose is cast loose."[34] Two men leaned back as they hauled the turtle up a sloping gangplank leading out of the kraal. The turtles were then lined up on the dock in preparation for butchering. Schroeder described the butchering in some detail:

Turtles that are used for canning purposes are slaughtered on the turtle dock. Each day during the greater part of the year five or six are killed at 3:30 p.m., at which time an inspector is present to see that the butchering is done in a sanitary manner. No turtles are killed until the desired number has been removed from the pens and laid about one foot apart on the dock. Then one person takes a sharp ax and strikes the head and four flippers off each turtle, going from one to the other with great rapidity. In each case the appendages are almost completely severed, allowing the animals to bleed freely. Immediately after the axman finishes, two men commence cutting away the plastron and then remove the entrails. During the operation sea water is thrown over the carcasses to wash away the blood and slime. The edible portions of the turtle are removed in four large pieces, each of which contains one of the flippers. The flesh is cut away from the carapace and thrown into a barrel of sea water, where it is thoroughly washed. It is then taken to the cannery, where it is hung on hooks and allowed to remain over night for use the next day. The following day a small portion of the meat may be sold for local consumption, but the greater part is used in preparing canned turtle soup.[35]

Schroeder noticed a certain amount of waste in turtle processing. He reported that "one prominent chef has stated that the carapace is one of the best parts of the green turtle for soup, but the Key West cannery disposes of this, the plastron, and the entrails by dumping them into the sea some distance from the shore."[36] The carcasses of the female turtles were stripped of their eggs; and both the white, mature eggs and the yellow, immature eggs were sold. Considered quite a delicacy, the yellow eggs fetched fifty cents per dozen, twice the price of the white eggs.

Just after Granday opened his cannery, at the peak of the Key West turtle industry, Brice felt compelled to report to Congress that "overfishing and the destruction of its eggs have greatly reduced [the green turtle's] abundance in [Florida], and the annual catch is now much less than formerly." His agents had learned much from Ralph M. Munroe, a naturalist and photographer who carefully observed Biscayne Bay's turtle fishing. Munroe told them that green turtles once bred in large numbers in the bights of the keys that form the eastern boundary of Biscayne Bay but now were very scarce. Brice concluded, "If their extermination is to be prevented they will have to be artificially propagated if the present indiscriminate methods are continued."[37]

Brice told Congress that it was desirable to establish a research and fish culture station in Florida, given that much more needed to be known about the diverse marine species there. And for the species that were better known, their biology seemed to lend themselves to artificial propagation, especially the Florida sponge, to extend and replenish the natural sponge beds.[38] With respect to the green turtle fisheries, he found that "overfishing and the destruction of its eggs have greatly reduced its abundance in this State and the annual catch is now much less than formerly."[39]

When it came to recommendations to Congress, Brice did not equivocate. He wrote that if the marine resources of Florida—which supported so many people and which are "one of the chief attractions to the visitors who annually resort to this region for health and pleasure"—are to be preserved, "some legislation is necessary." To prevent the green turtle from sharing the impending fate of the alligator, which was "rapidly approaching extinction owing to often wanton killing during recent years," Brice's prescriptions were clear:

> The green turtle, one of the most valuable of the State's fishery products, needs protection to prevent its extermination. For a term of years, at least, the animal should be unmolested during the period when it seeks the shores to lay its eggs. There should be a minimum limit of weight for turtles that are taken to be shipped or sold locally, in order that the destruction of immature turtles may be prevented. The pernicious and destructive practice of gathering the eggs of this and the loggerhead turtle should be prohibited.[40]

But neither the US Congress nor the state of Florida heeded these recommendations. Perhaps because the fishery occurred in diverse regions of Florida, there were too many explanations for the green turtle's decline. Some people thought the steamboats had frightened the turtles away from the Indian River Lagoon. Others attributed the decline to the severe freeze of 1894–95, to the construction of jetties at the lagoon's passes, or to excessive egg collecting and overfishing.[41] The Key West turtle trade was still strong, and the state of Florida had higher priorities in managing its coastal industries. No member of Congress was of a mind to protect the green turtle, nor to direct the US president to negotiate a treaty for its conservation.

In January 1898, the governor of Florida invited delegates from the states and the US Fish Commission to attend a National Fishery Congress in Tampa, Florida. Isaac Kibbe attended and presented a paper on oysters and oyster culture in Texas, the fishery that was now of concern to many states in the Gulf region. The one paper on green turtles was written by Ralph Munroe but presented by Hugh M. Smith of the US Fish Commission. Munroe's warning echoed Brice's report. He wrote of the green turtle: "Being, as it is, a nutritious delicacy, it is quite time that its habits, reproduction, and methods of capture should be looked into before its enforced classification with the extinct reptiles, even if this should be an event far distant; and it might well be worth our time and attention to reduce, by cultivation and protection, the present rather prohibitive price of a valuable food."[42]

Munroe's call for turtle cultivation was heeded—but for another highly prized, edible turtle, not the green turtle. In the early 1900s, the US Fish Commission opened a culture facility for the northern diamond-back terrapin (*Malaclemys terrapin*) at its research laboratory in Beaufort, North Carolina, with the object of developing methods that could be used in small-scale commercial farms throughout the southeast coast.

Two decades later, a young man in Savannah, Georgia, worked briefly at one of these terrapin farms. His name was Archie Carr. When he later became a biology professor at the University of Florida and wrote a treatise on the turtles of the United States, he reflected on the idea of terrapin farming with amusement. He described the "utopian state" that the US Fish Commission foresaw resulting from its "paternal aim . . . to carry on investigations that would place the terrapin culture on such a firm basis that the demands of the market might be supplied, the natural populations might be saved from extinction, and numbers of people in the tidewater areas of the United States might be provided with an extra source of food and income."[43]

Carr noted that when Hugh M. Smith, who had read Ralph Munroe's paper at the National Fishery Congress in 1898, later became commissioner, he readily endorsed the idea of turtle culture and foresaw a great future, writing tongue-in-cheek:

> Professional gourmands and confirmed epicures may soon be
> confronted with a question that to them at least will be a
> momentous one. If the choicest item in their repertoire of things
> aquatic becomes so common that the wayfaring man and other
> equally impecunious persons may easily purchase and consume it,
> will the diamond-back terrapin continue to enjoy its vogue? Will
> gastronomic fashion sanction the further use of the diamond-back
> as the scintillating gem in the diet of the elect? Will the millionaire
> wish to have this creature served on his table if the cost does not
> exceed that of a baked potato of the crop of 1918?[44]

But Carr found Smith's sanguinity regarding turtle culture misguided. Despite the commission's efforts, it seemed unlikely that small-scale farms would be feasible or popular or "that there will ever be a terrapin for every pot." The public had lost its taste for terrapin soup just as quickly as it had acquired it during the Gay Nineties, not because of Prohibition or the Great Depression, but owing to it being "an irrational fad and, like any fad, it passed."[45]

CHAPTER TWO

Turning Turtles on the Great Barrier Reef

B y 1895 the Texas and Florida green turtle fisheries were finished. The Key West cannery and the New York soup makers now depended entirely upon shipments of giant green turtles delivered to Florida by the Cayman Island schooners who caught the turtles on the Miskito Bank.

On the other side of the world, the settlers of Queensland, Australia, were just discovering an abundant local population of green turtles and the value of the "edible turtle" to the mother country. In the City of London, no mayoral banquet was complete without serving clear green turtle soup.[1] At aldermanic banquets and in the baronial halls, green turtle soup was served in lavish silver tureens made by the most sought-after silversmiths, some in the shape of a green turtle.[2] To the gentry, the green turtle had become the symbol of magnificence and abundance, of a civilization that had tamed the New World and gotten rich in the process. To the citizenry, green turtle soup and the long-robed aldermen who consumed it symbolized luxury at the expense of the common people. While Isaac Kibbe, the Texas commissioner of fish and oysters, was measuring green turtles lined up at the Fulton cannery for compliance with the state's twelve-inch rule, his counterparts in Brisbane, Queensland, were reporting with approval the expansion of the commercial green turtle fishery at Moreton Bay, made possible by the introduction of refrigeration.[3]

Exports were the core of the Queensland economy, and marine fish and wildlife were important parts of the trade, along with minerals from the dry country and the abundant grasses for pasturage in the wetter parts of the state. When a royal commission inquired into what ailed Queensland's principal commercial fisheries—for pearl shell and the sea cucumber called *bêche-de-mer*—the green turtle fishery still appeared very promising.[4] Merchants who were shipping green turtles to markets and soup purveyors in London invested in new facilities on the expectation of continued abundance of the turtle supply. The fishing industry had found the green turtles nesting on the tiny coral islands of the southern Great Barrier Reef. Adapting the method the Europeans used in the Caribbean, they captured every female turtle that came ashore by turning them on their backs and then collecting them in the hours and days that followed. The turtles were shipped live to small industrial ports in Queensland and often shipped on to Europe. A canning factory was built near the turtle nesting beach on the Barrier Reef's North-West Islet. During the breeding season, the factory canned them almost as fast as the turtle turners could catch them.

The Great Barrier Reef had been known to Europeans for over a century as a source of green turtles. Captain Cook first noted them during his 1770 voyage when he charted the reefs and claimed the eastern seaboard of Australia for the British Crown. By that time, European explorers and the French and English navies had already depleted the vast green turtle herds of the western Caribbean by visiting the Cayman Islands and taking on board a load of live green turtles to have fresh meat on the voyage home. The Jamaica law of 1770 banning the capture of turtles had had no effect. Like its Caribbean counterparts, Cook's *Endeavour* reported taking twenty-one large turtles in twenty-seven days when it stopped for repairs near the Endeavour River.[5]

In late-nineteenth-century Queensland, the sea turtles were still plentiful.[6] Early in the twentieth century, London's demand for turtles slackened with the onset of the European conflict. But after the war, demand for turtle soup resumed. Two new canneries were built on the Great Barrier Reef, in the Capricorn-Bunker islands group off the town of Gladstone, on the extreme southern coast of Queensland. The Barrier Reef Trading Company built a new canning works at North-West Islet. By 1924, the Australian Turtle Company had a cannery on Heron Island. When the nesting turtles were plentiful, trade was brisk, with catches averaging 2,500 green turtles per year. By 1925, a Queensland fisheries agent, Mr. V. Forrester, was advising his employers of the need for regulation.[7]

The green turtle fishery also attracted the attention of two curators at the Australian Museum. In 1926, after a three-week expedition to North-West Islet, Gilbert Whitley, the curator of fishes, and Anthony Musgrave, the museum's entomologist, published a chilling account of the turtling operations on the island in the museum's magazine.[8] Intended to be merely an account of the three species of sea turtles they encountered, the article

FIGURE 3 Boat owner, employees, and officials standing on a river boat looking at catch of green turtles before delivery to canneries, at Fitzroy River, Queensland.
Collection of the State Library of Queensland, negative number 13994.

told of turtle hunters patrolling the tiny island's beaches, turning over every female as soon as she had crawled onto the beach above the high tide line. The gasping giants were often left on their backs through the next day, baking in the sun while awaiting the boat that would take them, piled high on every inch of space on the deck, to the cannery.[9]

When the nesting season was a busy one, production at the cannery was high. Musgrave and Whitley reported that enough turtles were taken during the 1924–25 nesting season to produce 36,000 tins of turtle soup. Since a good day's catch of 22 to 25 turtles produced about 900 tins of soup, the 1924–25 season must have taken about 1,000 turtles. The curators questioned whether this was a sustainable fishery: "Each of the twenty-five odd turtles which are killed daily during the breeding season is the potential mother of about one hundred and fifty young so that unless drastic measures are taken, the species in the long run will become extinct."[10]

The curators' discussion of the possibility of extinction of an important part of the reef ecosystem appeared just as the public was recognizing the value of the Great Barrier Reef. Scientists and holidaymakers traveled there to marvel at the wondrous species and sights. In 1928, the Oxford Expedition to the Low Islands brought a group of British scientists, their spouses, and various assistants to the reef. They stayed an entire year, setting up a camp on North-West Islet, not decamping until July 1929. The expedition was sponsored by the Great Barrier Reef Committee, which had been formed in the 1920s for the purpose of investigating every scientific aspect of the coral reefs and islands. The expedition included ornithologists, marine biologists, geologists, and zoologists. Although most of the scientists came from abroad, one member of the expedition was a young Australian invertebrate zoologist named Frank McNeill. Another was a marine biologist, Frank W. Moorhouse, who worked for the Queensland inspector of fisheries.

The wondrous discoveries of the Low Islands expedition were shared with an international audience when, writing for the *National Geographic Magazine* in 1930, the Australian photographer and natural history author Charles Barrett described the trip. He told of scientists organizing a turtle derby, an event in which the lighter members of the group, usually the women, knelt on the backs of large green turtles that had been overturned the previous evening to keep them from returning to the sea. The women then "raced" to the sea trying to maintain their balance on the heaving turtles. The visitors also toured the turtle soup canneries operating on the islands. Barrett described a visit to a turtle soup factory on one of the coral isles, noting that in a recent season it processed 1,000 turtles into 36,000 cans of soup. He asked, "How many of those who enjoyed the soup at banquets or private dinner tables thought of atoll and lagoon, white beaches, and the hunters of the chelonian?"[11]

The Low Islands expedition was soon followed by a group of school teachers from New South Wales. People began to expect to see giant turtles when they visited the Great Barrier Reef, hoping to engage in the new sport of turtle riding.[12] Although this was not without its impact on the turtles, the sport and the tourism meant public sympathies were beginning to turn toward the green turtle, and the public wanted to see it alive and moving along the beaches and reefs of the popular new marine playground.

Frank Moorhouse returned to Heron Island soon after the expedition had packed up and departed. The inspector of fisheries had asked him to make further studies of the marine life and, in particular, to determine the potential of the *Trochus* marine snail for culture. Moorhouse dutifully completed this work but decided to extend his visit to make a more complete study of the green turtles he had seen during the Low Islands expedition's visit to the Capricorn-Bunker islands group. He ended up staying three and a half months, almost the entire breeding season, from October 1929 until February 1930. In his report to the Great Barrier Reef Committee, his comments on this exploited species' prospects were far from sanguine.

Due to the scale and pace of production at the Heron Island turtle factory, Moorhouse learned that the green turtles had become so scarce during the 1928–29 season that turtle-turning teams had to visit other islands in order to keep the cannery in active production. But the rate of exploitation led him to make an important inference about the green turtle's reproductive biology. Since the catch in each season had been virtually 100 percent, he reasoned that the entire breeding-age population must not return each year. Instead, only a portion of the herd returns in a given year, while the others waited some interval of years before returning to nest again. By marking the nesting females with tags throughout the season, he was able to observe that each female breeding that year returned up to seven times to lay a nest. With so many repeat nestings, it was apparent to Moorhouse that the breeding population at any one island was much smaller than the factories' owners assumed. Contrary to the turtle companies' advertisements in Brisbane newspapers, the green turtle supply was not inexhaustible. He wrote: "The idea now prevalent that there are thousands of turtles visiting any one island during the breeding season is quite erroneous and must be replaced by *a limited number of turtles make many visits to any one island during the breeding season.*"[13]

Rather than leave any doubt as to what needed to be done, Moorhouse took pains to recommend the exact nature of the regulations that should be adopted before the next breeding season if the turtle population was to survive the current level of export. Moorhouse noted that Queensland currently had no restrictions on the taking of turtles. Fishermen were thus free to take as many as they could, even taking turtles before they had laid their eggs: "Though the short-sightedness of killing the turtles before they had laid is admitted, even by the hunters themselves, this unwise practice is still followed. If it is continued, especially early in the breeding season, it must in the near future deplete our stock of turtles to such an extent as to wipe out this branch of our fishing industry; therefore a regulation should be framed in order to prevent the extermination of the turtle" (20).

It was essential to prevent the taking of turtles until all the animals that were breeding that season had had a chance to lay at least one nest. This, he estimated, would occur by the end of November. Therefore, he recommended a regulation "similar in form" to the following: "No person, south of latitude 17 degrees South, shall take, or offer for sale, and no person shall purchase, kill or attempt to export, between the dates of 30th September and 30th November of each year, any turtle of the kind known as Green Turtle (*Chelonia mydas*). Penalty £10 for each animal found in possession" (20).

Moorhouse emphasized that a closed season was "absolutely essential, for it is the only definite means of ensuring the laying of some of the eggs normally produced by the turtle" (22). He also suggested that the fishermen and factory authorities be required to plant the eggs removed from the slaughtered turtles, but he stopped short of proposing specific regulatory language for this or for a catch limit or minimum size, two common measures in fishery regulations: "Since it has been demonstrated that such action is advisable and profitable, then the yearly production of young turtles can be appreciably increased. It does not appear wise at the present juncture, owing to insufficient knowledge, to frame regulations to limit the number of turtles that shall be taken each season, or the size of animals permitted to be taken" (21).

Moorhouse would not go any further than his data allowed him to go in proposing regulations. He knew, for instance, that the egg-laying season went as late as May each year in the area north of latitude 17 degrees south (at approximately Cairns), including in the Torres Strait islands, but he cautioned that the breeding season in these areas should be scientifically delineated before a closed season was adopted.[14]

Perhaps sensing that Moorhouse's recommendations were well-supported by fresh field observations, Queensland's officials acted quickly to implement them.[15] By December 15, 1932, the Queensland Parliament had gazetted an Order in Council prohibiting the capture of green turtles for the first two months of the known breeding season everywhere south of latitude 17 degrees south.[16]

By then, however, the factories on both Heron Island and North-West Islet had already closed, one due to legal and financial problems and the other to the scarcity of turtles. The canning season that followed the year of Moorhouse's study had closed before the end of the breeding season for lack of turtles to process. By 1935, the Heron Island factory had been converted into a resort for reef and turtle tourism, and the transformation of the green turtle from marine food to friend in this part of Queensland was complete.[17]

Turtle fishermen continued to catch turtles in other places in October and November and in the cays after the closed season ended. They delivered their live catch to the meat works at Rockhampton to be made into turtle soup, and into frozen and dried turtle meat and bone, and shipped to Brisbane and overseas for the turtle soup lovers. But by 1939, Gladstone was no longer the center of the turtle fishery; catching was taking place in northern Queensland at nesting beaches on cays in the Torres Strait. The indigenous people living near these beaches started their own commercial turtle fishing companies, and high catch rates were reported by the Queensland fisheries officials by the early 1950s.

In January 1950, another zoologist who had been on the Oxford Expedition to the Low Islands gave his own, more personal impressions of the Queensland turtle industry. Frank McNeill had just completed a vacation at the Heron Island resort. When he disembarked on the pier at Gladstone Harbor, he and his companions were aghast at the sight of the giant green turtles, gasping for breath in the hot sun while waiting on the dock to be shipped to Brisbane.[18]

McNeill returned to Sydney to his post as curator of invertebrates at the Australian Museum and immediately began looking for information on the green turtle export business. He and P. D. F. Murray, a professor of zoology who was with him in Gladstone, found that during the egg-laying season, between twelve and eighteen turtles per week were being sent by rail from Gladstone to Brisbane. There, they were slaughtered and shipped as carcasses, along with the chilled beef exports, to England and Europe for the luxury food trade.[19] As an accredited affiliate of the Great Barrier Reef Committee in Brisbane, McNeill wasted no time in calling the committee's attention to the destruction of the green turtle by the preceding four decades of fishing. In his letter of protest, he noted that the fishery concentrated entirely on females, giving the species no time to recover between seasonal raids on the nesting beaches. He found that Queensland's two-month closed fishery season, adopted in 1932 following Frank Moorhouse's recommendation, was not being enforced.

Although a professional biologist, McNeill mounted as much an anticruelty campaign as a conservation protest, writing also to the Royal Society for the Prevention of Cruelty to Animals. He later wrote of the incident that had compelled him to look into the turtle fishery:

By chance some strangers were destined to come upon that deplorable scene. They were among a number of passengers who disembarked from the motor cruiser *Capre*—holidaymakers homeward bound to a southern State from coral-girt Heron Island in the Capricorn Group. Away from the tempering sea breeze, the general discomfort of the still heat caused an immediate and sympathetic reaction to the plight of the suffering animals. The newcomers watched resentfully while a miserable small stream of water from a hose was played on the captives. . . . Instead of alleviating the creatures' distress, it seemed only to aggravate their disablement.

They impotently responded by thrashing about with their flippers and struggling in a hopeless way to escape from their tormentors. Here was proof of an ill-considered and cruel exploitation—a practice calculated to endanger the very existence of a quaint edible marine reptile in one of its last world-strongholds. Cruelty of this kind has a way of continuing unabated until noticed by someone determined enough to take decisive action. . . . No time was lost in gathering convincing evidence for the strongest possible of protests.[20]

The Great Barrier Reef Committee made an inquiry into McNeill's protest in May 1950. By that time, McNeill and Murray had garnered publicity for their campaign in the newspapers. This, in turn, led the director of Queensland's tourist bureau to convey to the undersecretary for harbors and marine the numerous complaints it had received from other tourists visiting the southern Great Barrier Reef who were similarly shocked by the cruelty.[21] As a result, the committee's hearing focused as much on the cruelty of the trade as on its excessiveness, and parallels were drawn to the increasingly unpopular hunt for humpback whales in Queensland's waters. The value of the growing tourist trade was compared with the paltry income that the turtle fishery generated.

But the committee's concern was with how little of the scientific knowledge of the green turtle had been applied to management of the green turtle fishery. Members noted that despite Moorhouse's recommendations in his 1931 report, the breeding season in the northern section of the Great Barrier Reef had not been investigated in order to limit the impact of the fishery on nesting turtles. By unanimous recommendation, the committee urged the government to mount an investigation into the ecological and economic status of the green turtle along the Great Barrier Reef. Pending that study, the committee recommended that the government place the green turtle on the list of animals protected under state law.[22]

Within six months, the Queensland Parliament responded. The two-month closed season was rescinded and replaced with a new Order in Council which stated that the law "doth absolutely forbid the taking of any of the species of Turtle known as 'Green Turtle' (*Chelonia mydas*) or the eggs thereof in Queensland waters or on or from the foreshores of or lands abutting such waters."[23]

Three years after the new law was adopted, Queensland's director of native affairs, C. O'Leary, who was residing on Thursday Island, received an account of wasteful green turtle exploitation similar to the one McNeill delivered to the Great Barrier Reef Committee, this time on Bramble Cay, a primary nesting ground in the Torres Strait islands. Captain A. Mellor, the master of the passenger vessel *Melbidir*, told O'Leary that the owner of a fishing vessel he met at Thursday Island reported seeing four boats working for indigenous commercial fishing operations load fifty to sixty turtles each night. The boats were so full that as many as ten full-sized turtles died after being left behind when the fully loaded boats departed for the factories.[24] The fishing crew of the vessel *Wanderlust* were taking equally large numbers of turtles at nesting beaches on Palm Island. The vessels appeared to be operating under an exemption from Queensland fishery laws for aboriginal fishing.

In 1958, the ban on commercial turtle fishing was lifted in the northern Great Barrier Reef in response to lobbying by commercial fishermen. The Queensland Parliament repealed its 1950 order and adopted a new Order in Council, of September 4, 1958, allowing green turtles to be caught in Queensland waters north of latitude 15 degrees south. The fishing vessel *Trader Horn*, whose owner and master had reported the exploitation of the Bramble Cay nesting beach to Captain Mellor in 1953, caught approximately 1,200 green turtles in January and February 1959, in the midst of the breeding season. This appears to have been the only use of the 1958 northern exemption, however, as the venture was not profitable.[25] Ten years later, the Queensland Parliament repealed the northern exemption and adopted a third Order in Council designating all sea turtles as protected species, having recognized that the green turtle was more valuable as a tourist attraction than as a fishery.[26]

One side effect of the development of the Great Barrier Reef as a tourist destination was the disturbance to nesting green turtles brought about by the popular sport of turtle riding at the resorts that cropped up on the Great Barrier Reef, particularly those at Heron, Masthead, and Lady Musgrave Islands.[27] As we have already seen, this activity had been going on since at least the Oxford Expedition of 1928–29.[28] Musgrave and Whitley reported riding turtles during the visit of the Royal Zoological Society of New South Wales to North-West Islet in 1926.[29]

Charles Barrett, whose 1930 article and photographs in *National Geographic* brought international attention to the Great Barrier Reef, described the process as involving much more than a spontaneous hopping aboard a sleeping turtle. Men were assigned the task of searching for nesting females the night before the "turtle derby" and turning them on their backs so they could be ridden the following morning.[30]

Organized derbies soon followed and became a tourist attraction in their own right. The Queensland government tourism board received complaints from some visitors who did not find the treatment of the turtles amusing, especially when the turtles were left on their backs for hours in bright daylight. The Queensland Society for the Prevention of Cruelty to Animals also received letters from appalled guests and in turn complained to the Queensland Department of Harbor and Marine in

FIGURE 4 Vacationers and green turtles on Heron Island, Great Barrier Reef, Australia, 1938.
Collection of the State Library of Queensland, negative number 65648.

1944. The practice was not outlawed for at least another twenty years, however, the Order in Council of 1950 having outlawed only the commercial taking of green turtles. Turtles confined to the beach for the sport were eventually released and therefore did not fall under this prohibition.[31]

CHAPTER THREE

The Turtle Islands of Sarawak

While the epicures of the European capitals were acquiring a taste for green turtle soup and thousands of live turtles were exported from the New World, people in the colonial hinterlands prized the green turtle's eggs as a delicacy. To feed this taste, eggs were systematically harvested from the major turtle rookeries and marketed. To ensure maximum revenue from this resource, colonial governments in places such as Malaya, Ceylon, and Borneo sold exclusive concessions to egg collectors for specific sections of the nesting beaches.

The nesting population that was most affected by these egg harvests was on the Turtle Islands of Sarawak, off the northwestern coast of Borneo. These three tiny coral islands—Talang Talang Besar, Talang Talang Kechil, and Satang Besar—lie off the mouth of the Simitar River and not far from Sarawak's capital city of Kuching. Eggs were systematically collected from these islands from as early as 1839, just before the establishment of the Brookes dynasty, the rulers known as the "White Rajahs."

The Brookes dynasty began in 1841 when the British baronet James Brooke helped quell a rebellion against the sultanate of Brunei and was rewarded by being made rajah of Sarawak. Brooke's nephew, Sir Charles Brooke, became the second rajah. He built a natural history museum in Kuching, on the recommendation of the English naturalist Alfred Russel Wallace, co-discoverer of natural selection independently of Darwin, who had spent two fruitful years of research in Borneo and environs. In 1946, Vyner Brooke, the son of Sir Charles and the last White Rajah, celebrated 100 years of Brooke rule in an enlightened manner. He approved a new constitution for Sarawak, abrogating his absolute powers in favor of a governance structure that would allow the British colony to move toward democratic self-government. Vyner Brooke also adopted a new turtle law.

The Turtle Islands of Sarawak were the nesting grounds for one of the largest populations of *Chelonia mydas* in all of Southeast Asia. The eggs from the Turtle Islands were the center of a small industry that began after James Brooke's suppression of piracy, with the help of the British Navy. Malay businessmen hired the men who did the actual egg collecting and then sold the eggs to merchants in Kuching and the coastal towns of southwestern Borneo. The nesting turtles had fascinated Rajah James Brooke. He wrote detailed observations in his diary of the scene in 1839 when he revisited the islands: "Morning calm. In the afternoon got under way and anchored again near the islands of Talang Talang. . . . The Bandar of the place came off in his canoe to make us welcome. He is a young man sent by the Rajah Muda Hassim to collect turtles' eggs, which abound in this vicinity, especially on the larger island. The turtles are never molested, for fear of their deserting the spot; and their eggs, to the amount of five or six thousand, are collected every morning, and forwarded at intervals to Sarawak as articles of food."[1]

When he visited the islands again three years later while pursuing pirates, James Brooke saw as many as 100 turtles come ashore each night in June and July, perhaps laying as many as 20,000 eggs on a good night. Twenty to forty men lived on the islands. In their watching hut, they waited for nesting females to come ashore. When a turtle had finished covering her nest, the watchers marked the spot with a flag. Brooke noted that the next morning, they "purposely spared some nests" when the eggs were dug up, dried in the sun, crated, and sold to wholesale dealers in Sarawak's market towns. When Brooke became rajah, he sought to preserve this practice. His penal code made it illegal to kill any sea turtle, meeting such destruction with a heavy fine.[2]

The second rajah, Sir Charles Brooke, also had ideas about what was best for the Turtle Islands. In 1875, Sir Charles broke the monopoly that Rajah Muda Hassim held over egg collecting and awarded the right to the leading Malay *datus*, or chiefs, of Sarawak. The datus agreed to rotate control over the egg collecting on an annual basis, each keeping the proceeds from their year of control. Nesting took place all year, but with a definite peak at the end of the northeast monsoon season. This season was met with celebrations, called the *Semah*, after which no one was allowed to land on the islands for three days to ensure that the turtle fertility shrines built during the ceremonies were not disturbed.[3]

Watchers hired by the Malay datus lived on each island. Their job was to mark the location of each freshly laid nest with a flag. Another crew would dig them up the next morning and send them by boat to Kuching.[4] The second crew recorded how many eggs were collected, a number that varied between three and four million per year.[5] The turtles themselves were never eaten; if the European expatriates in Sarawak wanted turtle soup for their dinner parties, it had to be served from a tin. Only in Singapore could one purchase fresh turtle meat and fat.

Edward Banks's Theory about Turtles

Edward Banks, a Welsh anthropologist and naturalist, was the Sarawak Museum's curator from 1925 to 1945. He considered it to be "one of the best jobs in the Far East" and used the position to study many things about Sarawak, including its fascinating green turtles and the industry built up around their mass egg-laying. By Banks's time, the Malay chiefs' exclusive egg concessions had been passed down under Islamic rules of inheritance into many hands, and there was no central control over the egg collecting.[6] But Banks was able to find records of the number of eggs that had been collected from the islands and sold. From these he tried to piece together a better picture of the turtles, the size of the population, its frequency of nesting, and the conditions that affected nesting productivity.

Banks had a theory to explain why the numbers of eggs laid fluctuated widely from year to year. While the average "bag" was two million eggs, the total could vary by a million each way, with good years making up for bad years. He thought it likely that the two million eggs were the product of 5,000 to 10,000 female turtles. He speculated that their variable production was due to the intensity of the wet monsoon, which disrupted breeding aggregations. The Malay owners of the turtle egg concessions told Banks that the low years' numbers were due to reductions in the number of turtles by fishermen or poachers, or to inadequate prayer meetings, or both. Banks tried to explain to the Malay head turtle-keeper the likely correlation of heavy monsoons and low laying seasons. The keeper rejected Banks's theory. It was contrary to what he had seen with his own eyes since the 1890s. Two years later, however, after another below-average season, the same keeper castigated Banks for taking so long in coming around to the monsoon explanation for low laying numbers.[7]

Banks wondered how the turtle population could survive the very high rate of egg collecting that the Malay chiefs required and whether some sort of intervention was needed. "On the whole it seems as if the natural fluctuations of the turtles are due to weather conditions and not to any decline in population," he wrote, but it is "almost incredible that there should be no such decline, for every year almost all the eggs are taken and few if any left for hatching on the islands. . . . And as it really seems unlikely that allowing the eggs to hatch and the young to escape will assist the population to increase very rapidly, it would appear that some artificial help is needed." But it was not clear what the appropriate means of providing that artificial help would be, given that all previous attempts at fencing in or building enclosures for the young turtles had failed. The one possibility would be to build a sanctuary on Satang Besar on the site of the former leper settlement, where a concrete dam from a reservoir was still in place and which was close enough to the water to be useful. He suggested that sea water could be pumped by a windmill or machine into the enclosure and the young turtles fed coconut products from the adjacent plantation for about six months, "until no longer a prey to the predaceous fishes when returning to the open sea."[8]

Despite the high mortality of hatchlings that he saw and the near-total take of the eggs, Banks took comfort in the fact that the Malays were "almost fanatical" about protecting the adult turtles. The green turtle eggs were one of the few commodities Malay communities had for trade. Thus, they never thought about killing these turtles or selling them in Singapore. The adult turtles weighed 200 to 400 pounds, so Banks estimated they took about ten years to mature. "With a breeding stock of five to ten thousand females, valued at about £4 in pre-war days, it might have been thought the Malays would have disposed of some of them; but [they] would have none of it, a decision which I rather share on finding that after the shell, fat and cartilage have been removed, only about one-third of this vast quantity is edible meat."[9]

Not long after Banks published his observations in the *Sarawak Museum Journal* in 1937, Rajah Vyner Brooke bought the egg rights back from the Malay chiefs' families. He directed his legislative council to proclaim a new law that would put the museum curator in charge of the Turtle Islands.

Chapter 40 of the Laws of Sarawak declared that it was the exclusive right of the government to take the turtles and eggs from within the territorial waters of Sarawak. A Turtle Board of Management would oversee the industry as a government business. The ordinance, and its counterparts in the states on the Malay Peninsula, reflected the political influence of the British, displacing the notion that rights to the turtles' eggs were a royal prerogative of the sultans. The laws put turtle exploitation under strict control to ensure a source of government revenue, along with Western-style taxation systems. The special Sarawak twist was the Turtle Board. The board would consist of several high-ranking people whose job was to ensure that proceeds from the sale of the one million eggs each year would be distributed appropriately. The museum curator would oversee day-to-day operations on the Turtle Islands. Although the islands were privately owned, under Chapter 40 the turtles were, in essence, held in trust for the people, with the profits from the sale of their eggs accruing to Malay charities and mosques.[10]

A New Curator for Sarawak

By the time the Japanese invaded Sarawak on December 25, 1941, Sir Vyner Brooke had already evacuated, and the Japanese interred Edward Banks along with the rest of the expatriate community. The new turtle ordinance sat on the shelf while Japanese bomber pilots used rocks near the Turtle Islands for target practice, and the occupying troops raided the islands for turtle meat frequently. After the war, the curator would have to restore the islands and the egg collecting industry if the Turtle Trust was to operate. Although the collections of Sarawak Museum were preserved, the Japanese had left the Turtle Islands in a shambles.[11]

With the end of the war, Banks decided not to resume his post as museum curator, and the British polymath Tom Harrisson was chosen to replace him. Harrisson had been to Borneo twice before and knew quite a bit about its diverse flora and fauna as well as its many peoples. In the 1930s, Edward Banks had suggested to Rajah Vyner Brooke that he invite the British zoologist Charles Elton to bring an expedition to explore the biological riches of the Sarawak highlands. Elton turned to Tom Harrisson, who had organized successful Oxford-Cambridge expeditions to Lapland and St. Kilda, in the Outer Hebrides, and had distinguished himself as a talented ornithologist.[12] Harrisson raised the money for the expedition to Borneo and picked its members, among whom was his friend Eddie Shackleton, son of the heroic Antarctic explorer Sir Ernest Shackleton.

Harrisson had arrived in Sarawak in 1932, and it was immediately apparent that the difference in his style and that of Banks, who was in charge of the expedition, was like the proverbial "chalk and cheese." Banks was a quiet and dedicated naturalist with a deep respect for colonial administrative protocol. Harrisson was a brilliant, heavy-drinking know-it-all with boundless energy, who adopted the local living style whenever he was on expedition in remote places, especially if it involved stimulating beverages and intimate relations with women.[13] When Banks reached the end of his patience, he complained to Charles Elton about the young men's behavior, especially the brash and undisciplined Harrisson. But it was Harrisson who published an account of the Oxford expedition to Sarawak in 1938.[14]

Despite his reputation as an *enfant terrible*, Harrisson's knowledge of Borneo and its peoples was by 1945 well regarded in the British government. When he joined the air force during World War II, the military tapped Harrisson to train an Australian special operations team and parachute into the Japanese-occupied highlands of Borneo. Their job was to gather intelligence, but they ended up leading the tribal people in guerilla warfare against the Japanese forces that were fighting the Allied landings.[15]

Harrisson distinguished himself in this service as a fearless and inventive organizer, and he had been reluctant to leave the people he had worked so hard to protect. He was especially fond of the Kelabit, an indigenous people of the Borneo highlands whom he had gotten to know during the 1932 expedition. The Kelabit had aided Harrisson's team immeasurably by telling them of Japanese troop movements and by killing and capturing soldiers with blowguns and poison darts. Along the way, they revived the infamous practice of head-hunting that the second rajah had taken pains to suppress. After the special operation, Harrisson had stayed on as a civil officer in the Kelabit district for a time. So when the war was over, and he heard that Banks would not be returning to his job in Kuching, Harrisson applied for the job and got it.

Harrisson had great hopes as he took up his post as the first postwar curator of the Sarawak Museum. He liked the fact that he would be government ethnologist as well as curator. He hoped to study Sarawak's diverse peoples and cultures and build upon the notoriety he gained before the war from his best-selling book, *Savage Civilization*, about living with the people of the New Hebrides after an earlier Oxford expedition. He had no idea that the rajah's turtle ordinance of 1941 required him to administer the turtle egg industry. He hoped to achieve many things for himself and for Sarawak, but had little thought that one of these things would be to pioneer research on green turtles and become a staunch advocate for their conservation.

Flirting with Turtle Science

When confronted with the mass of statistics on the numbers of eggs collected and sold each season, Harrisson realized how little evidence lay behind Banks's estimate that the Turtle Islands' nesting population was between 5,000 and 10,000. Without a more certain number, Harrisson would not know how many eggs should be taken for sale and how many eggs allowed to hatch to ensure a continued supply. He also couldn't know the amount of revenue that would be generated from the egg sales. He needed a better idea of the number of mature females that made up the nesting population, the number of nests they laid in a given season, and most importantly, the number of years they spent at sea before returning to nest again. He could count the eggs they laid in each nest and derive the total number of eggs in a season. But the rest required data that he did not yet have.[16]

Harrisson considered himself something of an expert in nature conservation. When Britain faced similar questions in 1930 about its bird populations, he had organized several large-scale censuses, often collaborating with Britain's most eminent zoologists and rising academic stars. It was the height of the bird conservation movement; and Harrisson, using rings, had marked and counted the great-crested grebe, the birds of Lundy Island, and the birds and mammals of St. Kilda.[17] While planning the grebe census in 1930, Harrisson became friends with pioneers of ecology and ornithology, including Charles Elton, Julian Huxley, and Max Nicholson, and many younger men who would go on to found leading conservation organizations.[18] When he looked at the eggs statistics Banks and the Malay chiefs had compiled, he thought they too showed a serious decline in the number of nesting females. He could not be sure, however, unless he had a better idea than Banks's of the total population size. Could sea turtles be marked with rings like grebes?

While back in London on his first home leave in May 1949, Harrisson visited the British Museum of Natural History where Hampton W. Parker, the keeper of zoology, helped him find the literature on *Chelonia mydas*. Harrisson soon realized that only a small number of marking studies had been done on sea turtles—one in the Danish West Indies and another on the Great Barrier Reef. He wrote right away to the Great Barrier Reef Committee for a copy of Frank Moorhouse's report on his Heron Island investigations of 1929–30 and for as much of the literature as he could find.

When he read Moorhouse's report, Harrisson realized that the green turtle populations of the Great Barrier Reef also were heavily exploited, but with a difference. The canneries on Australia's coral islands were killing adult females, while the Sarawak turtle industry only collected eggs. Moorhouse reported that on Heron Island, females were often killed before having a chance to lay their eggs. Upon his recommendation, the Queensland government required the canneries on the reef to plant the eggs of the female turtles that were killed. Harrisson could see, however, that the numbers of eggs that hatched from these artificial nests were not encouraging. He thought perhaps more careful handling and natural egg-laying could ensure a greater hatch rate. Moorhouse had used a tag attached by wire looped through holes drilled into the rear edge of the carapace to mark the adult females. This allowed him to count the number of times they returned to nest and the number of eggs each one laid during the season.

Moorhouse's study inspired Harrisson to do similar but more extensive work on his turtle population. Harrisson was confident he could improve on both Moorhouse's artificial nesting and marking methods. Moorhouse had studied only one nesting season; he retired soon after compiling the Great Barrier Reef expedition reports. No one else appeared to be studying the green turtles of the Great Barrier Reef or anywhere else in the South Pacific. The field was wide open to someone with access to the kind of nesting population Harrisson had in his care under the rajah's turtle ordinance. He could control all activities affecting turtles on the Turtle Islands, rebuild the population, and make the industry profitable again, generating revenues and reputation for the museum. He resolved to "go at this strongly at Sarawak."[19] He would begin by redoing the study of Edward Banks, his predecessor and former nemesis.

Harrisson corresponded with Paul Deraniyagala, the director of the National Museum of Ceylon, another former British outpost with an important nesting colony of sea turtles. Deraniyagala had published a number of sea turtle taxonomic studies in the *Ceylon Journal of Science*, a publication of the Ceylon national museum, as well in the prestigious *Proceedings of the Zoological Society of London* and the British science journal *Nature*.[20] The Sarawak Museum also had a journal, which Harrisson planned to revive and thereby build the museum's reputation for excellent work on the cultural and natural resources of Sarawak.

Once he was back in Sarawak and on the Turtle Islands, Harrisson revived his passion for wildlife biology and ethology, experimenting with different hatching conditions, watching for evidence of light orientation, making observations on growth rates, and even looking for evidence of learning in the young turtles. He increased the number of nests that were allowed to hatch. In Banks's time, only a few nests were allowed to hatch out, and the hatchlings were allowed to run into the sea. Since the turtle watchers were already digging up the nests, he followed Moorhouse's example and devised a hatchery system.[21] From 1947 onward, he had the workers transplant a certain number of eggs to artificial nests and then hand-raise the hatchlings until they were large enough to fend for themselves. To give them a fighting chance at life, Harrisson used the museum's launch to take the young turtles out beyond the reef and its hungry denizens to where they could swim to open waters.[22]

Every time Harrisson, who was a compulsive writer, had interesting observations, he wrote them up under the heading "Notes on the Edible Turtle, *Chelonia mydas*," for the *Sarawak Museum Journal*. But his first note on the Turtle Islands appeared in 1950 in the *Royal Asiatic Society Journal, Malayan Branch*.[23] It was a history of the Turtle Islands' egg industry. Harrisson gave an account of the *Semah*, the festival that took place at the end of the monsoon season, demonstrating his ethnographic skills by recording a multicultural celebration of the turtles' anticipated productivity. Harrisson's writing style was conversational rather than

FIGURE 5 Sarawak Turtle Board's green turtle hatchery at Talang Talang Besar, British Borneo, July 1953, where John Hendrickson applied cow-ear tags to the foreflippers of nesting green turtles for the first time and where, three years later, Barbara and Tom Harrisson first witnessed a flipper-tagged female come ashore to nest.
Collection of the Sarawak Museum, courtesy of Christine Horn.

scientific, using wit and common words like "babies" and "ladies" to make the information accessible to those in his readership who might be inspired to send money to the museum to support further studies. Along the way, he made important observations despite his somewhat unorthodox methods, like raising green turtles in the bathtub of his guest bedroom. He initiated the use of the term "frenzy" to describe the behavior of newly hatched turtles and hypothesized that this behavior allowed them to get rapidly out to the open ocean, away from predators to where food might be available.[24]

One of the first articles Harrisson published in the Sarawak Museum's journal, after reviewing the egg collection records for 1950, was an

FIGURE 6 Tom Harrisson onboard the Sarawak Museum's launch studying green turtle hatchlings before releasing them seaward of the turtle islands, October 1961.

Collection of the Sarawak Museum, courtesy of Christine Horn.

update of Banks's 1937 study of the green turtle's breeding habits on the Turtle Islands. It was a response of sorts to Frank Moorhouse's 1933 study of nesting behavior at Heron Island on the Great Barrier Reef. Harrisson sent a shorter version as a letter to the British journal *Nature*, announcing to the world that research on this important species was under way in Sarawak. Convinced that Banks's study had been overlooked because of contradictions in the data, Harrisson reported nesting figures obtained from the staff he had stationed on the Turtle Islands as soon as he returned from London. He knew the figures were accurate because, since 1947, the Malay datus—chiefs—no longer controlled the three islands; he did.

He could report that in 1950 over 2,350,000 eggs were recorded (precisely 2,357,664), including eggs replanted for hatching. While peak nesting occurred in the summer, there was no question that eggs were laid every month of the year, even in the height of the monsoon season, perhaps owing to the islands' location so close to the equator, at two degrees north latitude. The monthly totals ranged from as few as 24,000 in January to as many as 500,000 in August. Contrary to Moorhouse's findings in 1933, which to that time was the most extensive green turtle study, nests in Sarawak had an average incubation period of fifty-two days, with a slower rate during the monsoon, but still much shorter than the sixty-five to seventy-two days Moorhouse had observed.[25]

Harrisson concluded the *Nature* letter as he often did the *Museum Journal* notes, by inviting correspondence and collaborations with other investigators: "There are interesting possibilities for comparative work over the enormous range of this common (though decreasing) species, the habits of which lend themselves to exact observation and statistical checking. We here would welcome any research co-operation or co-ordination."[26]

A Clever Solution from a Farm Magazine

John R. Hendrickson was a young American zoologist in 1951 when he accepted his first faculty job at the University of Malaya in Singapore. Unlike Tom Harrisson, Hendrickson had completed his undergraduate studies in biology and after the war had gone on to complete a doctorate, studying under the noted herpetologist Robert C. Stebbins at the University of California at Berkeley. His dissertation was on the slender salamander genus *Batrachoseps*, but Hendrickson soon realized that the Malay peninsula and the island of Borneo offered many other, very interesting animals to study.

A regular reader of the journal *Nature*, Hendrickson had found his teaching position in Singapore through one of its job advertisements.[27] When he saw Tom Harrisson's letter on breeding habits of the green turtle in the February 1952 issue of *Nature*, he began to think there might be nesting beaches in peninsular Malaya where he could study these intriguing reptiles. Soon after, he received an invitation from Harrisson, on behalf of the Sarawak Turtle Board, to visit the Turtle Islands. He quickly accepted. Hendrickson was especially pleased that Harrisson had written to his department chair, Professor Richard D. Purchon, seeking his young colleague's help. Hendrickson was anxious to get back into the field and work on challenging biological questions.

When Hendrickson made his first visit to the Turtle Islands in 1952, there was almost too much to do. The most pressing need was for a method to mark the turtles when they came ashore to nest. It was the only way researchers would be able to determine how often a green turtle breeds and whether the females return to the same beach from which they had hatched. Hendrickson doubted any of the marking methods the egg workers were using would produce useful information on the size of the population. Harrisson, the boisterous and bossy curator, as executive officer of the Turtle Board kept a research bungalow on the main turtle island of Telang Telang Besar but made only infrequent appearances. He encouraged Hendrickson to experiment with different marking methods until he found one that would serve the purpose.[28]

There were two problems with the tagging method Moorhouse had used on the Great Barrier Reef. It required the turtles to be turned over onto their backs so that the workers could drill holes in the carapace and attach the tag with wire. Once the tags were attached to the turtles, they were not particularly durable, often coming back corroded and illegible.[29] One day while he was back at the university in Singapore, Hendrickson was perusing an agriculture journal when he noticed an advertisement for a self-piercing, self-clinching tag that dairy farmers could apply to the ears of their cattle with patented applicator pliers. At the same time he heard that the Institute for Medical Research in Kuala Lumpur was having a set of sinks made from Monel metal, a nickel-copper alloy that does not rust. He had an idea.[30] He would order a custom-made set of Monel cow-ear tags to apply to the turtles of Sarawak. If applied to a tough section of turtle hide somewhere on the foreflipper close to the body, the tag would not act as a fishing lure or otherwise hurt the turtle's chances of returning successfully to the nesting beach.[31] The tag makers back in Kentucky were only too happy to oblige. When the bulk order of specially fabricated tags arrived in Sarawak that summer, Hendrickson set about doing some serious tagging.[32]

The new Monel tags worked well and could be applied quite easily. To ensure that the research team got the most information possible, each tag was embossed on one side with a code letter indicating on which island the turtle was tagged, followed by a number. On the other side each tag bore the words "Sarawak Museum" and "Reward."[33] From March to October 1953, Hendrickson and his student research assistant

FIGURE 7 Sarawak Museum turtle flipper tag used by John Hendrickson and Tom Harrisson at Sarawak Turtle Islands, British Borneo.
Courtesy of Carla Kishinami.

tagged as many turtles as they could on all three of the Turtle Islands. He made plans to return and tag more turtles during the next years' peak nesting seasons. If the turtles had a two- to three-year interval between breeding, he could tag the entire population of adult females and help Harrisson get the numbers he needed.

Field Notes and Firsts

The curious curator was not there when Hendrickson returned to the Turtle Islands in early 1953 to apply the new tags. Harrisson had left the previous December for six weeks in the Borneo interior doing other museum-related research. When Harrisson had not returned by February 1953, as scheduled, no one seemed too concerned. He had a tendency to change plans and wander off to other places without notifying his staff.[34]

Hendrickson kept up the tagging program, managing to tag 1,514 females on Talang Talang Besar by April 1953. He also set about making numerous other observations of the turtles. He measured and weighed them on the beach, devising ingenious methods for hoisting them on

FIGURE 8 John Hendrickson examining turtle tracks at a beach on Sarawak
Turtle Islands, British Borneo.
Courtesy of Carla Kishinami.

scales used for weighing copra, the white meat of the coconut palms that
grew on the islands. Hendrickson was a tireless and thorough field biologist. He filled countless notebooks as he observed the turtles' energetic
mating behavior in the water, followed hatchlings by swimming and by
paddling next to them in a canoe, and excavated every nest after the
hatchlings emerged, counting every egg that had failed to hatch to get a
sense of the natural hatch rate. He planned to use these notes later when
he was back at the university to develop a complete picture of the observ-
able portions of the green turtle's life cycle.[35]

Hendrickson knew that if the tags applied early in the nesting season
stayed on, they would learn how many times a turtle returned in one
breeding season and get a more accurate picture of the size of the nesting population. He hoped the tagging would also result in longer-term
data that could shed light on whether these turtles made large-scale migrations and used a homing mechanism to find their way back to the

Turtle Islands. Knowing these things would make it possible to conserve and manage the turtle egg industry. At present, the evidence for homing and migration was lacking.[36]

Managing the industry would also require knowing the green turtle's growth rate and the age at which it reached reproductive maturity. Hendrickson measured the carapace length and width in 200 nesting females. He then compared the ratios of the carapace length to width with ratios measured in turtles raised in captivity in Ceylon, on the Seychelles, on Heron Island, and on the Turtle Islands, whose ages were known. From these data he made "cautious speculations" about the probable growth rate, concluding that the Sarawak turtles matured in less than six years but more than four.[37]

One day, while Hendrickson was busy taking these measurements, Harrisson reappeared out of the blue. The delay in his return had been due to a crash of the plane sent to pick him up from his research site in the Borneo highlands, the plane having crashed as it attempted to take off with him on board. Still ailing from the scrub typhus he had contracted while working in the highlands, Harrisson had walked down from the mountains and along the coast. When Harrisson found out that the entire egg-collecting staff had been engaged in Hendrickson's research and that Hendrickson had asked the Turtle Board to hire additional local people to help, Harrisson was furious.[38] Hendrickson was conducting expensive and intrusive experiments that he, as curator, had not authorized, with additional staff he had not hired. The place looked like a laboratory instead of an egg-collecting business. All that hoisting and prodding probably offended the Malay staff and dignitaries who managed the Trust on behalf of the Malay charities.

And where were the tagged turtles? None had returned. Harrisson thought they had probably quit the place because of the intrusive and bothersome experiments. But after Hendrickson explained why the work was necessary for managing the egg harvest, Harrisson convinced the Malay members of the Turtle Board to allow the tagging to continue, stretching a smaller number of tags in 1954 over nine months. But after only a couple months of tagging in 1955, he bowed to the concerns of the Malays and suspended the tagging indefinitely.[39] Because no one knew how long it took before a female returned for another nesting season, there was no way Harrisson could reassure the Turtle Board that the wait would be rewarded. Besides, Harrisson, the experienced bird-bander, doubted whether the tagged turtles would return. He was skeptical of the tagging because Hendrickson, with all of his scientific methods, had made it into such a big production and had probably disturbed the turtles enough to prevent their return.

With the Sarawak tagging program suspended, Hendrickson went instead to San Francisco, to present a paper on the new turtle tag at the annual meeting of the American Society of Ichthyologists and Herpetologists and to visit his PhD advisor, Robert Stebbins, at Berkeley, leaving his detailed field notes on the tagging program with Harrisson.[40] One of the meeting's attendees listened with particular interest to Hendrickson's paper. David Caldwell was a graduate student in zoology at the University of Florida and in the middle of his field work for a study of the migratory habits of the Atlantic green turtle in Florida. He knew immediately that his professor, the principal investigator of the study, Dr. Archie Carr, would be very excited to learn of the new tag when Caldwell returned to Gainesville.[41]

At the Sarawak Museum, Harrisson had all but given up on obtaining any information from the new tag by July 1956, when he and his new wife, Barbara, spent a working honeymoon on Telang Telang Besar. On July 4, they watched in amazement as a flipper-tagged female green turtle came ashore to nest. From the field notes, they determined that Hendrickson had applied the Monel tag to her foreflipper on July 30, 1953. The new tag had been a success after all.

Harrisson took little time in reporting this event to the world—and taking credit for it. He wrote a note for the *Sarawak Museum Journal* and composed another letter to *Nature*.[42] In his long and rambling museum journal note, "Tagging Turtles (and Why)," he explained how the success had happened. He told of realizing that in order to meet his obligations to the Turtle Board, he needed to know how many females made up the egg-laying population. But he had difficulties in finding a satisfactory method until the fold-over tag "was devised with the *well-helped help* of Dr. John Hendrickson, Department of Zoology, University of Malaya," who was the invited guest of the Turtle Board.[43] The letter to *Nature*, which would obviously reach a larger audience, made no mention of Hendrickson.

Some months earlier, Richard Purchon, Hendrickson's department chair, had told his younger colleague that Harrisson was publishing notes in the *Sarawak Museum Journal* based on Hendrickson's field notes without acknowledgment. Purchon encouraged Hendrickson to confront Harrisson and insist that Harrisson give him and the University of Malaya due credit for the work.[44] When Hendrickson spoke to Harrisson, however, Tom insisted he was entitled to use the field notes. He was the one who had instigated the research back in 1947, had gotten the money from the Turtle Board, and had issued the invitation to Hendrickson.[45]

But the young scientist's complaint bothered Harrisson enough that while editing the proofs for the museum journal, he added the begrudging acknowledgment of Hendrickson's contribution.[46] Harrisson realized that if he was going to build the museum's reputation for research on and conservation of Borneo's fauna, he would need the enterprise and expertise of zoologists like Hendrickson. He would also need help with the Turtle Islands' nesting turtles if the population size was as low as he now suspected. The tagging returns were showing that the Sarawak turtles nested as many as seven times in a season. Given an average clutch size of 100 eggs, the total number of females in the nesting population was likely to be around 5,000. This was far less than the 10,000 females he thought might be responsible for laying the one to two million eggs per season. If the 4,000 female turtles that had been tagged in 1953 to 1956 never returned because of the experiments, he might have added a significant drop in the nesting population to the gradual decline he detected from the egg statistics.[47] But before he would invite Hendrickson or anyone else to work with the Sarawak turtles again, he would wait and see if Hendrickson's obsessive concern for statistical rigor paid off in useful results publishable in a good journal. In the meantime, he would enjoy taking credit for the clever new turtle tag.

Harrisson always sent reprints of his articles to everyone he thought might help his turtle studies or Sarawak and its museum.[48] One letter in particular had just such an effect. He sent the second note, on copulation behavior of the "edible turtle," to Archie Carr, the noted zoologist at the University of Florida, who had just completed a taxonomic handbook on the turtles of the United States, Canada, and Baja California. When Carr wrote back thanking Harrisson for the "very interesting paper," he asked him for a copy of the first note, which was on breeding habits.

Carr also asked Harrisson for his advice on using the cow-ear tag as a means of marking sea turtles, which he believed Harrisson had described in a recent paper at the annual meeting of the American Society of Ichthyologists and Herpetologists, which Carr had missed. Carr wrote: "In my migration study I want something that will stay on for a long time. . . . Will you please tell me, at your earliest convenience, where you have found the most satisfactory place for the fin-tag to be. I expect to tag about a thousand turtles in Costa Rica this July and I hate to think of the tag falling off. I'll be very grateful for any advice you may care to give me."[49]

Harrisson read Carr's letter as soon as he returned to the museum from a trip to Europe. By this time it was already the height of the field season in Costa Rica. Harrisson apologized to Carr for the delay in replying and gave him the advice he sought. There was just one problem. Harrisson had not delivered a paper at the 1955 meeting and had not actually done any tagging. Hendrickson had done both. Harrisson nevertheless consulted the field notes that Hendrickson had left with him.[50] He enclosed a sample tag in his reply to Carr and wrote, "We apply them on the rear side of the forward flipper quite close to the main body. . . . So far we have had one recovery. . . . I am far from satisfied about this method and should be glad to hear about the results of your experiments."[51]

When Carr wrote back and told him he had adopted the Monel cow-ear tag for tagging green turtles at Tortuguero and had applied them himself in the last days of the 1955 field season, Harrisson knew he was onto something.[52] He now realized the global significance of what he and Barbara had seen when the first female turtle, which Hendrickson had marked on the flipper with the Monel tag, came ashore to nest after an absence of three years. Like the great-crested grebes he had banded in the 1930s, the green turtle was beginning to look like a candidate for a more vigorous conservation intervention. Harrisson suspected that the growing consumption of green turtle soup he had witnessed while in London on home leave was killing off his population of green turtles. Certainly, Carr had sounded the alarm, in his taxonomic handbook on turtles, about the serial extirpation of turtle populations in Bermuda and the Caribbean.[53] Harrisson thought of mobilizing to the green turtle's cause his friends and colleagues in England who were active in international wildlife conservation.

Harrisson actually had no idea of what was happening to his population when they were not breeding at the Turtle Islands. With few if any tag returns from beyond the coast of Sarawak, he had no proof that his turtles were being captured and shipped abroad to be made into green turtle soup. Maybe the old Malay chiefs who had argued with Edward Banks were right. Perhaps fishermen and poachers were killing the turtles for meat and shipping it to the soup-eating capitals of Europe. Banks was correct in attributing the nesting fluctuations to the monsoon but wrong in not noticing the overall downward trend.

Harrisson decided to revisit the social science work he had done in the 1930s when he and others established Mass Observation in England, an organization that conducted the first large-scale public opinion survey. He would use the data and methods he had pioneered to change the public's attitude about green turtles and its taste for green turtle soup.

CHAPTER FOUR

The Gifted Navigators

When Archie Carr rose to give the keynote address to the meeting of the American Institute of Biological Sciences, he had a different group of turtle islands on his mind. Christopher Columbus called them Las Tortugas. The islands were teeming with sea turtles that looked "like little rocks," Columbus wrote in his journal in May 1503, during his fourth voyage to the New World. The islands he described would later come to be known as the Cayman Islands, and their vast herds, or "fleets," of breeding green turtles would supply European voyagers, vessels, and colonies for the next 300 years.[1]

It was September 6, 1954, and Carr had been thinking about the turtles of the Caribbean for more than a decade. On that day in Gainesville, he talked of *Chelonia*, how its size and abundance reflected its straightforward ecology, and how its taste and ease of capture supported the opening up of the Caribbean and probably the entire New World. "All early activity in the New World tropics—exploration, colonization, buccaneering and the maneuvering of naval squadrons—was in some way dependent on the turtle," Carr said. Grazing all day in vast herds on submarine grasses, "they grew fat and numerous and succulent, and in every way a blessing." But it was a blessing that was short-lived. The last fleets of *Chelonia* were passing; he believed fervently that unless the turtle was "effectively protected it may soon be extirpated as a breeding resident of American waters." The green turtle would go the way of the buffalo, but without the notice and drama that accompanied that extermination. The green turtle was simply "too good to last."[2]

Six Big Questions

Carr had come to believe that the Atlantic green turtle was destined for premature extinction two years earlier while writing his taxonomic treatise on turtles, *Handbook of Turtles*. In writing the notes on the Atlantic green turtle, he had found historical sources on Bermuda and Jamaica describing the extirpation of one rookery after another as the New World's turtles fed the new colonies and the visiting navies. He cited Samuel Garman, whose 1884 report on the reptiles of Bermuda described how the Bermuda Assembly in 1620 was so concerned about overexploitation that it passed "An Act Agaynst the Killinge of ouer Young Tortoyses," prohibiting the killing of turtle less than eighteen inches in diameter from the waters around Bermuda to a distance of five leagues.[3]

In the 1950s, green turtles were still being exported in large numbers from the Caribbean to the United States. Carr had seen the impact of this trade with his own eyes as he scouted the Caribbean for sea turtle nesting beaches. Turtle soup canneries still operated in the United States and Europe, and Queensland was still exporting turtles caught by Aboriginal fishing companies.[4] But the Cayman Islands no longer had the magnificent breeding fleet that Columbus had seen. To stay in the turtle trade once their own herd was gone, the Cayman Islanders had built a fleet of turtle fishing schooners to find seagrass pastures where the green turtle could still be hunted. For the last hundred years or so, the Cayman schooners specialized in catching the green turtles residing on the Miskito Banks, some 350 miles away, off the coast of Nicaragua in the western Caribbean.

Since long before the war, Carr told his audience of biologists, a green turtle cannery in Key West, Florida, has done a brisk trade with the schooners of Grand Cayman.[5] Carr hastened to say that the Miskito Bank turtle fishery was not the most serious drain on the greens of the Caribbean. That was happening where the turtles gathered and came ashore for breeding in Costa Rica. The capture for export from a place called the Turtle Bogue was a relatively new fishery. Because it took adult breeding turtles, it was likely to be the last straw for *Chelonia*.[6]

Carr told his audience that the extirpation of the Atlantic green turtle could be prevented, but the knowledge needed to do so was lacking. The will was there, but the science was not. He noted that "at this point our procedure is blocked by an astonishing ignorance of the biology of the animal. Most of the countries and peoples with a stake in the Caribbean littoral would be in sympathy with the idea of saving the green turtle.

But once persuaded of the necessity of doing something they embarrass you by insisting that you tell them exactly what to do."[7]

But turtle fishing people did not lack this knowledge. Carr said that fishermen throughout the Caribbean and Florida believe that the green turtle is migratory, but "nowhere in the canons of zoology is there a shred of what could be called scientific evidence to prove it." Especially knowledgeable were the Cayman schooner captains, who had an unparalleled knowledge of the turtles and a stake in their survival. It was their belief that the turtles left the Miskito Bank in the spring and made their way south to Turtle Bogue. The captains not only believed in these mass migratory feats, they were convinced that the turtles found their way by virtue of a strong homing instinct. Their proof was that individually recognizable turtles could be found sleeping beneath the same rock night after night, returning to the same spot after a day spent grazing in seagrass pastures that were miles away from their favorite sleeping holes. The fishermen were convinced also by the story of the storm of 1924, one of several stories Carr had collected in his wanderings throughout the Caribbean looking for turtle nesting grounds. A large male green turtle with distinctive shark-bite scars on his flippers was captured on the Miskito Bank, branded with the schooner's initials, and sent to the cannery in Key West, which was soon hit by a hurricane that October. The next season, the Caymanian captain who had caught the turtle saw him again sleeping under the same rock where he had been captured the previous season. Carr concluded that the story had to be true, for it lacked any sentimentality added by the teller for effect. The captain had caught the turtle a second time and sold it again to the same dealer in Key West.[8]

Carr asked aloud what could account for the popularity of the Turtle Bogue as the primary nesting beach for the Caribbean green turtle. Predators and hunters abounded there. It was an exposed shore with heavy surf and no obvious landmarks to make it easy to find by the herds of turtles coming from both the south and the north to breed. It was on the beach at Turtle Bogue, known to Costa Ricans as Tortuguero, that Carr intended to find out how *Chelonia* accomplished its long-range feat of migration. But at this twenty-four mile stretch of black beach, a law was in effect that had the exact opposite intent of the 1620 Bermuda ordinance. The Costa Rica law created an orderly and highly efficient

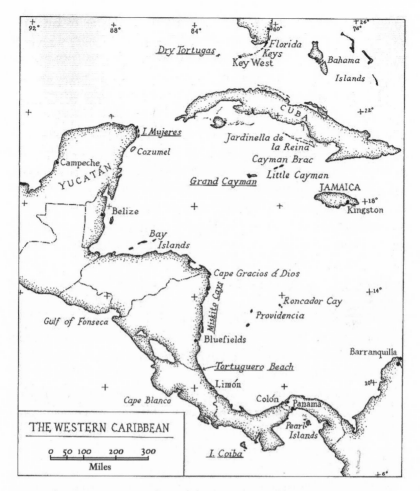

Geographer James Parsons' map of the western Caribbean, 1961.
From James J. Parsons, *The Green Turtle and Man*, courtesy of David J. Parsons.

process for taking all the turtles, and Carr knew it had to be changed before he could start the tagging program that would establish the scientific fact that green turtles were mass migrators.

Costa Rica's law gave the government ownership of the first mile of shore land extending inland from the surf zone. Every year it sold leases to mile-long sections of the beach at public auction. The highest bidders, the *contratistas*, were given the right to take all the turtles that came ashore on that section of the beach. They hired local people to patrol the beach all night and turn on its back every turtle that came ashore to nest. By law these turtle-turners, the *veladores*, were supposed to wait until the female had laid her nest. Carr saw that this rule was always broken because it cut down on the number that could be captured. The veladores hoped to turn thirty to forty turtles in a night and get thirty-eight cents for each. And besides, there was no one there to know otherwise. The veladores turned each turtle onto her back to await collection and built a shelter of coco thatch so she would survive the wait for the launch sent by law every two days to collect the turtles. Carr had seen turned turtles die by the dozens when bad weather delayed the collecting launch from Limón. The veladores floated those that survived out to the launch across the surf by tying chunks of balsa wood to their fins, and they were gathered in by canoes working for the launch.

Carr described the economics at work in the turtle fishery, and told his listeners that sooner or later they would change in the direction of doom for the turtles. Once the demand and price were high enough, the launches would be more reliable and the veladores even more efficient. The increasing numbers of settlers moving to the Costa Rica coast meant it was unlikely the turtles could establish alternative rookeries to get away from this population drain.

This was a grim prospect but also one that gave him hope. For the very same reason that the green turtle was vulnerable—its habit of mass breeding—the number of beaches that needed to be protected was small. It was indeed possible "to bring back the fleets Columbus found."

> It is not often that we are offered a set of circumstances so promising; a one-item feeder with its pastures undamaged, vast in extent and used by no other animal; a species attuned to building and thriving in dense populations and yet flexible enough to proliferate and scatter in dilute colonies; above all, a depleted species, the cause of whose depletion is clear and surely possible to remedy. There is still a skeleton breeding stock and the best of the remaining nesting shores are the least cluttered by man. Group action by the governments concerned would surely save Chelonia and build unity and strength in the Caribbean by raising the yield of the sea to the people around it. In the field of live-resource management, it is not often you can hope for so much.[9]

It was a remarkable speech. Carr displayed the gifts he would soon become known for: an ability to combine natural history observations, scientific curiosity, and a lively, uncluttered prose to convey the mystery and magnificence of the green turtle.[10] Also evident were his abiding concern for survival of this species and recognition that both science and law were necessary to achieve it. Efforts to negotiate an agreement among the countries of the Caribbean for the protection and restoration of green turtles had little hope as long as key aspects of their biology are not known.[11]

Carr was still optimistic that science would overcome political and economic considerations. Two years earlier he had written that if we could answer just six questions, "the specific information which in this connection seems most necessary would be available." Those questions were

1. Where are the principal remaining breeding areas located?
2. What is the breeding season at each of the nesting localities?
3. How many times does a female lay during one year and how many eggs constitute her annual total? Is laying an annual occurrence, or do the turtles lay on alternate years as has been suggested?
4. Where do the hatchling and post-hatchling stages go and what do they do prior to attaining "chicken turtle" size—five pounds or so in weight?
5. What diurnal, seasonal, and developmental migrations, if any, take place, and to what extent might an increase in the population in one area be expected to replenish that in another? Are the Caribbean and South Atlantic stocks genetically isolated from each other? These points could readily be clarified by marking experiments.
6. What is the volume of the annual turtle and egg takes throughout the range of the animal?

The good news was that these questions could be answered: "If adequate solutions to these problems could be obtained—and they await only a proper investigation—it seems probable that the green turtle could not only be saved from virtual extermination but might even be encouraged to regain something approaching its primitive range and abundance."[12]

This was in 1954, eight years before Rachel Carson's *Silent Spring* sounded the alarm over the impending extinction of songbirds due to humankind's profligate use of the pesticide DDT. The green turtle had found a champion who was as determined to understand its biology as he was to save it from extinction. Carr was not, however, falling victim to scientism—the optimistic belief that science could solve all societal problems including biological conservation. He could see clearly the economic forces at work that would require a countervailing political and legal pressure. He believed that these pressures could only arise when there was more biological and ecological knowledge about the green turtle.[13]

The Research Station at Tortuguero

The biggest of Carr's six big questions was whether the green turtle is a long-range migrator. To answer it, Carr needed reliable tags, a team of people to attach them to lots of turtles, and a stretch of beach where the turtles would be unmolested by the veladores. Any study of the Caribbean green turtle would have to focus on the rookery at Tortuguero, Costa Rica, and it was there that he would set up his field station. While he sought grant money from the National Science Foundation, Carr asked the Costa Rican Ministry of Agriculture to allot a two-mile section of beach adjacent to the station where the veladores could not turn the turtles. Carr then hired the veladores displaced from their own industry to work on his, turning the turtles for tag and release rather than capture and export.

After three summers of reconnaissance, it was finally time to start the work. In July 1955, the research station was established at facilities in Tortuguero that a banana company put at Carr's disposal. Once Carr got the tagging program started, his friend and former professor Leonard Giovannoli, with help from the veladores, would tag as many turtles as possible.[14] If the tags remained on the turtles during their suspected migration, the reward offered on the tag would motivate the turtle's catcher to return the tag to Carr with information on where and when it was caught. With the data thus received, he could test the hypotheses about the mass, long-range migration he had gleaned from conversations with the turtle captains of Cayman Islands.[15]

Giovannoli was a master tagger. He managed to tag fifty overturned turtles in one night, using Monel metal wire to attach inscribed oval plates to the overhanging back edge of their carapaces.[16] But the shell tags eventually came off; they could not withstand the scuffle and abrasion of the males' mating behavior. Carr had been testing different tag types in a pilot study with his graduate student David Caldwell, using green turtles caught off the Cedar Keys of Florida in the Gulf of Mexico.[17] While Carr went to Tortuguero to open the field station, Caldwell attended the annual meeting of the American Ichthyologists and Herpetologists in San Francisco.

When Carr got back to Gainesville, Caldwell told him he had heard a paper about researchers in Borneo using a narrow strip of Monel metal to tag green turtles, bending it over the flesh of the front flipper. Carr immediately ordered a supply of tags from the company in Kentucky. The supply of flipper tags arrived just before the end of the first field season. Carr rushed it down to Tortuguero, and he and Giovannoli applied forty of the new tags in four days before closing the station for the season. It was the Monel cow-ear tag that John Hendrickson had designed for the Turtle Islands to replace the copper plates that were like the ones used by Frank Moorhouse on the Great Barrier Reef.[18]

Carr did not have to wait two or three years to conclude that the flipper tag was a great success. The tags stayed on and demonstrated that females returning to nest within the season could reemerge at almost the exact spot where they had previously emerged, showing "some sort of orientation accomplishment." Carr saw these as "feats of great stature" because it was unlikely the turtles had loitered off the exposed shoreline for the three to four weeks they were absent.[19]

All of Carr's tags offered a five dollar reward for their return, and this seemed to get immediate results. Fishermen returned tags from ten of the turtles he and Giovannoli had tagged during the first field season. Seven of the ten came from the Miskito Bank, the area some 300 miles northwest of Tortuguero, off the coast of Nicaragua, where the Cayman Islands turtle schooners operated. This was the first positive evidence of long-range migration, at least of individual turtles if not en masse. Carr was very pleased with the new tagging method and decided to report the initial results in *American Museum Novitates*, a publication of the American Museum of Natural History, in New York City.[20] Between 1956 and 1966, Carr would publish a total of six notes in *Novitates*, including papers with his student Harold Hirth and his research assistant Larry Ogren. James A. Oliver, the director of the American

FIGURE 9 Applying cow-ear tag on turtle flipper at Tortuguero.
Special Collections, George A. Smathers Libraries, University of Florida, Gainesville, courtesy of Carr Family Trust.

Museum of Natural History and a close friend of Carr's, assisted Carr's migration studies.[21]

A Marvel of Nature

One Cayman turtle captain in particular provided the bulk of Carr's evidence of the green turtle's migration. Carr had developed his hypothesis that green turtles were long-range migrators with a strong homing instinct after hearing Captain Allie Ebanks's stories of recapturing turtles branded with his schooner's initials in the same place after they had escaped from the pens at Key West. Now, the same Captain Ebanks, on

FIGURE 10 Dragging turtle to scales. From left to right: Harold Hirth, Larry Ogren, Leo Martinez, and Archie Carr.
Special Collections, George A. Smathers Libraries, University of Florida, Gainesville, courtesy of Carr Family Trust.

the *A. M. Adams*, the schooner that he captained for Thompson Enterprises in Key West, would return a tag from a turtle that had been flipper-tagged at Tortuguero at the end of the 1955 season and was caught on March 3, 1956, at Southeast Rock, south of Miskito Cay. The fisherman, who had used his own marking method to assert ownership of a sea turtle, would provide evidence of the superiority of the new flipper tag.[22]

Not only was Carr looking for evidence of mass migration; he was also looking for scientific information that would convince the Caribbean nations to cooperate to save the green turtle from certain extinction. The green turtle in American waters could not be saved without "protection from international laws, and the laws will have to be based on an understanding of the life history of the animal." But there was no conclusive evidence of the link between the catching of turtles in the western Caribbean and the populations in American waters. The Cay-manian captains believed turtles had a homing mechanism, but there was no scientific evidence to support these beliefs. Without the science, the green turtle would not get the protective laws it needed.[23]

Carr recounted some of the anecdotes upon which the Cayman schooner captains based their belief in a homing ability in his book *The Windward Road*. The best story was the one Carr heard from Allie Ebanks, from 1942. Captain Allie had let one of the men who caught the turtles from the *A. M. Adams*'s small catboats select and mark five green turtles and send them home to his family via the *Lydia Wilson*, which had finished its trip and was returning to Grand Cayman the next day. Two weeks later, the same catboat fisherman recaught one of the marked turtles that he had put on the *Lydia Wilson*, at the very same rock where he had caught it the first time. This convinced Ebanks and the crew of the entire fleet that the *Wilson* had been lost in the storm they had just endured, freeing the turtles to return to their sleeping rock. But several days later Ebanks learned from another passing schooner that the *Wilson* was safe at home in Grand Cayman, 350 miles away, and that the storm had flooded the turtle pens at Georgetown where the turtles were being kept. "So the story changed quickly from tragedy to just a marvel of nature." The animal had not only returned; it had followed the shortest and best route, using either a special sense or its ordinary senses in "some clever way." Carr was convinced that the turtles that made up the breeding fleet at Tortuguero were migrants from Miskito Bank. When he received the tags from Allie Ebanks in 1956, he was sure of it.[24]

A Misunderstanding in Gainesville

While Carr's manuscript on the 1955 field season at Tortuguero was in press for the *American Museum Novitates*, he received a manuscript from John Hendrickson based on Hendrickson's investigations at the Turtle Islands of Sarawak. Tom Harrisson's letter describing the results of the Sarawak tagging project had just appeared in the December 29, 1956 issue of *Nature*. Hendrickson decided it was time he published a paper from his field notes; he was the one who had made the detailed observations that Harrisson was relying on. If he could not get back to the Turtle Islands to continue the work, he would at least contribute what he knew to this important species and discuss what he thought were the best approaches to its utilization and management.[25]

Hendrickson had seen Carr's 1954 grantee report of his reconnaissance of the turtle beaches in the Caribbean and knew that Carr was also studying the behavioral ecology of the green turtle. In this short note Carr concluded that in American waters, the green turtle seemed to be "in a dangerous state of depletion" but that it was also "most peculiarly amenable to conservation manipulations."[26] What Hendrickson had not yet seen was Carr's first paper in the *Novitates*. In it, Carr and his student David Caldwell reported on their pilot studies of migration and homing by the green turtles that Florida fishermen caught in a seasonal fishery off the west coast of Florida. The paper described testing three types of tags inscribed with a reward and return address: two sizes of Monel metal disks attached by wire to the carapace, and one 2-inch cow-ear tag attached by a special pincer to the fore flipper. All three were being tested for durability on captive turtles at the aquarium in Fort Walton Beach, and they were likely to use the flipper tag for the remainder of the study. Carr and Caldwell's paper, however, mistakenly attributed the idea of using the cow-ear flipper tag to a paper presented by Tom Harrisson at the 1955 annual meeting of the American Society of Ichthyologists and Herpetologists.[27] It was Hendrickson who had given the paper, and it was Hendrickson who had come up with the idea.

Hendrickson asked Carr for an unsparing review of his Sarawak manuscript, telling him he had decided to stop revising the paper and send it off for review. After eighteen months, it was clear to him that he was not going to be invited back to the Turtle Islands to continue the study. He asked Carr to send it on to Robert Stebbins at Berkeley when he was finished and told him Stebbins would repay him for the postage. His plan was to submit it to *Proceedings of the Royal Society of London*, but he invited Carr to suggest other outlets.

Hendrickson apologized for not seeing Carr's paper with Caldwell on the Florida greens but promised to cite it in the final manuscript. He assured Carr that his falling out with Harrisson was not over who had thought up the cow-ear tag. He wrote: "It doesn't much matter at all about who first had the idea. . . . I am not sure I didn't get the notion from someone. . . . [As] you will see, the turtles Harrisson works on are the same ones [that I did the] major part of my work with. . . . I tagged a lot while I was working there, and . . . extra tags and pliers were left on the islands."[28] Harrisson's letter in *Nature* had mentioned that 4,000 turtles had been tagged; Hendrickson himself had tagged 2,404 turtles and had overseen the tagging of another 316. He concluded that Harrisson had continued with the tagging after he had left the islands.

He acknowledged that his falling out with Harrisson was to the detriment of the turtles and to science but thought it in bad taste to discuss the details in a letter. "I feel that you need some explanation of the background," he said, and asked if Carr would be willing to let Hendrickson acknowledge his assistance in reading the manuscript. "It may be important for you to preserve good relations with Harrisson, and any implication that you have helped me might not go down well with Tom." He offered to have a set made of his turtle photographs from Sarawak in gratitude for Carr's help and to deliver them in person in the coming summer.

Carr wrote back two months later. He apologized for the delay, saying he had been busy with the start of the new academic term and had wanted to read Hendrickson's paper closely. He had assumed he would have lots of corrections to make, given the "concurrent evolving of your project and two progress reports of my own [the two papers in *Novitates*], but it turns out I have held you up for nothing." He liked the paper very much and told Hendrickson that he could indeed say that Carr had read it and thought it was "a damn fine job." There was only one bone he wanted to pick: Hendrickson's paper had shed no additional light on the issue of migration. He urged his young colleague to keep an open mind:

> Your material was quantitatively so much more imposing than ours
> has been that I was hoping to get reflections in the movements of
> your turtles that would reinforce or help interpret my own case
> histories. You and we are working on widely separated populations,
> and probably on genetically different creatures, but they are obvi-
> ously not very different and I suggest that you not rule out migration
> (not that you have) as a possibly important aspect of the life history
> of your turtle. It is of course conceivable that, as you seem inclined to
> believe, your rookeries are all recruited from among local residents.
> With the lack of any spermatophyte grazing flats in the area and
> with no big feeding aggregations known from anywhere around I

would remain very receptive to the possibility that your nesting turtles come in there, possibly in schools or waves that overlap too much to show up in your frequency data, from over a wide territory. We have shown a six hundred mile spread in recruitment for our rookery here and I am sure we have only begun. If you stay out there I surely hope you can figure out a way to tackle this aspect of the problem. That is what interests me most about the work here, and once we have established long-range migration as a fact there will be spectacular possibilities for studies in orientation and navigation mechanisms.[29]

But Carr would soon have more data from his own tagging studies and the strong evidence he sought for the long-range migrations the turtle captains had told him of. The tag returns from the station at Tortuguero were generating results, thanks in no small part to the idea Hendrickson had had while reading an American farming magazine in Singapore.

CHAPTER FIVE

The Geography of Turtle Soup

Another scholar who was drawn to the green turtle was James J. Parsons. Neither a zoologist nor a curator but a geographer from the University of California at Berkeley, Parsons believed a geographer's place was in the field. He was, in fact, an economic geographer with a regional focus on Latin America. By the mid-1950s, Parsons had studied a range of topics in American agriculture and natural resource–based industries. Plants were his particular passion; his detailed assessment of the introduction of African grasses to the Americas became a classic in the field. While Parsons was studying the geography of the Miskito pine savanna, the southernmost stand of pine on the American continent, he came across Archie Carr's 1953 book, *High Jungles and Low*, and its vivid description of forests along the part of the eastern Nicaraguan coast known as the Miskito Shore.[1] Carr had written the book after returning to Florida from his four-year stint teaching at the Pan-American School of Agriculture in Honduras.[2]

Parsons visited and wrote about the English-speaking settlements of the Miskito Shore and the offshore islands of San Andrés and Providencia.[3] The trade that connected these islands with the Miskito coast introduced Parsons to a different species of the American tropics, whose interesting geography had not yet been told. Switching from flora to fauna, he decided to document the geography of exploitation of the green turtle.

The Green Turtle and Man

When he began to explore what was known about the green turtle, Parsons soon realized that Archie Carr had also begun to focus on the green turtle, finding in the literature accounts of Carr's research in Florida and Costa Rica, alongside the studies reported by Tom Harrisson and John Hendrickson from Borneo and Malaya. Between these three scientists, the green turtle's nesting and migration habits were well on the way to being understood. But the history of exploitation of the green turtle had never been pulled together in one account. Parsons decided he would enlarge upon the growing biological knowledge with an account of man's use of the green turtle, illustrating the impact of this use with a map of all the nesting and feeding areas where turtles and their eggs had been hunted, the turtle places that no longer existed and the few that remained. He would explore the different cultural attitudes toward eating turtle flesh and eggs and the history of international trade in green turtles. He would also document where governments had tried to manage the exploitation with licenses and prohibitions, following the unsuccessful 1620 Bermuda ordinance, which Carr had quoted in his *Handbook of Turtles*.

To get all this information, Parsons corresponded with the turtle scientists, with the many geography colleagues he had around the world, and with trade officials in Asia, Africa, and the Caribbean, peppering his correspondents with questions about turtle localities and markets and the practices of local peoples. He asked John Hendrickson about marine turtle localities and cultural practices in Southeast Asia, noting Hendrickson's account of the origins of the Muslim attitudes toward eating turtles that appeared in his 1958 paper in the *Proceedings of the Zoological Society of London*. To fill a gap in information on the turtles of Siam, Hendrickson told Parsons to try to obtain a paper that a Thai naval officer had presented at the November 1957 Pacific Science Congress in Bangkok.[4]

Parsons began his 1962 book *The Green Turtle and Man* with an essay on the "the world's most valuable reptile." After a brief review of what was then known of the green turtle's biology, he described the cultural attitudes toward the meat and eggs. The European view of the turtle's value had evolved over time. When first discovered, the green turtle was valued only as an antidote to scurvy but soon became a staple of the West Indian plantation and colonial diet. By the Victorian era green turtle soup was a prestige food enjoyed only by the wealthy, "a symbol of Victorian opulence" occupying the same class of fare as the oyster. In other cultures, eating turtle flesh was a cultural taboo, although the eggs were collected and consumed in vast amounts.

Parsons compiled dozens of historical and modern accounts of human uses of green turtles, documenting how the green became the focus of a global international trade. He went far beyond the US Fish Commission's reports and the account Carr gave in his *Handbook of Turtles* chapter on the Atlantic green turtle to find a lethal connection to the shores he was most familiar with. The English explorer William Dampier brought Miskito Indians from the Caribbean on his voyages to the East Indies in the late 1600s to serve as "strikers" to capture green turtles for the ships' larders.[5] When Parsons could find no modern records of turtles on historical nesting beaches he realized that human consumption had wiped clean countless rookeries.

Parsons was particularly struck by how the demand for green turtle soup was wiping out turtle rookeries. He found historical sources indicating that the English gentry learned to like turtle soup while stationed in the Caribbean to oversee the sugar plantations and colonies. By shipping live turtles home to London, they helped the City of London's elites develop an epicurean taste for green turtle soup, the demand for which led to a hefty cross-Atlantic trade in turtles. The finest restaurants all offered green turtle soup, and from the mid-1800s the soup was produced on a larger scale by soup manufacturers. When Lusty's of London began to can turtle products in 1870, demand for turtles kicked into high gear. By 1878, over 15,000 live turtles were being shipped across the Atlantic, most of them caught by the Cayman turtle schooners and shipped via Jamaica to the capitals in Europe.

The soup manufacturers imported green turtles from all over the world, including Ascension Island in the South Atlantic and Kenya on the Indian Ocean. But the primary source was the West Indies, with the Cayman turtle schooners supplying the largest number. Royal Navy officers often brought live turtles from Ascension Island and the West Indies to London for Lusty's to make into soup for presentation to the lords of the admiralty.[6] As the official purveyor of real turtle soup to the royal household, Lusty's began to import frozen carcasses and then bought dried calipee to save on shipping costs. When it made soup for the lord mayor's banquet for the City of London, Parsons wrote that Lusty's produced "a rich gelatinous, clear soup that will set in a natural jelly even before it is completely cold and will almost 'stick the lips together' when eaten." The preparation took up to five days and required over 2,000 pounds of turtle, the equivalent of ten whole turtles.[7]

Lusty's was joined by soup manufacturers in many cities on the Continent. In Germany, Lacroix of Frankfurt-am-Main, began producing soup in 1921, exporting much of its product to the United States. In America, Moore & Company of Newark, New Jersey, was the leading soup manufacturer and purchaser of green turtles from the Caribbean, shipped from Tampa and Key West, Florida.[8]

Parsons found trade statistics indicating that in the early 1960s, imports by the European canneries were down to about 1,200 frozen carcasses, most of which came from East Africa, and thirty tons of dried calipee. The chief fisheries officer in Nairobi, Kenya, told Parsons that the green turtle had just been declared royal game, and the fishery was now subject to a system of permits. It was now illegal to turn a turtle on the beach or take one under twenty-four inches in size. Even with these restrictions, the Kenyan official estimated that 2,000 would be taken in 1959, and of those, between 1,200 and 1,500 would be exported to Europe.[9]

Culture and Attitudes

Parsons was interested in documenting the different cultural attitudes about eating turtles and their eggs. He described the Turtle Trust ordinance in effect in Sarawak and provided an account of the *Semah* fertility ceremonies that he had gotten from Tom Harrisson's article in the *Sarawak Museum Journal*.[10] Parsons concluded that where the eating of turtle meat was culturally accepted, the turtle nesting beaches were more likely to be depleted. He cited John Hendrickson's assessment that protecting turtles as egg providers was a more effective conservation strategy than prohibiting the taking of eggs and that the Muslim practice of not eating turtle flesh might be the only thing saving Southeast Asian sea turtles from extirpation. He noted Hendrickson's conclusion that the geographic extent of the practice of avoiding consumption of sea turtle flesh—a trait shared by residents of the coasts of the Malay Peninsula, Burma, and Thailand—suggested it was a pre-Muslim attitude. Among modern Thais, it was "unthinkable" to eat a sea turtle.[11]

When Parsons finished the manuscript of *The Green Turtle and Man*, he wrote to Carr and asked if Carr would review it and advise him on where it should be published. Carr agreed immediately, acknowledging it would be an important asset to his migration studies. After he read it, he wasted little time in recommending that his friend Lewis Haines, director of the University of Florida Press, look it over. Carr told Haines that the paper, like Parsons' previous work on San Andres and Providencia, was "sure to find enthusiastic approval in his own field" as well as among scholars in many other fields. Carr also suggested his friend Graham Nutting, director of the Carnegie Museum in Pittsburgh, as the manuscript's outside reader. He wrote Nutting, warning that the manuscript would be coming: "It will be in your role of geographer that you'll be sought. Without trying underhandedly to influence you—unduly—I want to say that this seems to me, as a turtle man, an extraordinarily fine paper, which, whatever its value as human geography, will at one stroke greatly enhance the background for studies of sea turtle migrations. As a herpetologist this ought to please you!" Nutting wrote back: "Your approval as the leading turtle expert will make it much simpler for me to concentrate on the geographic aspects. Certainly if publication will aid the beleaguered marine turtles in any way I would favor it."[12]

The Brotherhood of the Green Turtle

In the final chapter of his book, Parsons considered the uncertain future of the green turtle in light of the growth of human populations. Books on the impending extinction of a species were uncommon in the early 1960s, although Carr would soon sound the alarm in his 1963 book, *The Reptiles*. But to show that there was some room for optimism, Parsons described at some length the first active conservation program to benefit the green turtle, which had grown directly from Carr's research at Tortuguero. He recounted how a group of men in the publishing business who had read and admired *The Windward Road* responded to the peril Carr described in his final chapter, "The Passing of the Fleet," which was based on Carr's 1954 address to the American Institute of Biological Sciences. They had formed the Brotherhood of the Green Turtle in 1958, for the purpose of "restoring green turtles to their native waters, and insuring Winston Churchill his nightly cup of turtle soup."[13]

The seed for the formation of the Brotherhood was planted by William M. Pepper, Jr., publisher of the *Gainesville Sun*, a newspaper in the city where Carr's University of Florida is located. Pepper recommended that his friend Joshua B. Powers, a fellow member of the Inter-American Press Association, read *The Windward Road*, after Powers spoke of having admired Carr's *High Jungles and Low*. Powers followed Pepper's recommendation. He thought so much of the new book that he sent it out to twenty influential friends and told them they were now part of the Brotherhood of the Green Turtle, a team that would rescue the turtle from extinction. By 1959, after the group had introduced themselves to Carr and offered to help his research at Tortuguero, Powers visited Costa Rica and convinced the Costa Rican Ministry of Agriculture to close the beach to turtling. This would protect returning nesters from the veladores so the turtles could lay their nests, be tagged or have their tags read, and then return to the ocean to go back to their feeding grounds. Closing the beaches would also keep the veladores from killing the turtles carrying tracking equipment as well as turtles that might be held on the beach for use in the cognition and physiology studies.[14]

In 1959 the Brotherhood incorporated as the Caribbean Conservation Corporation (CCC). With Carr as its technical director, the corporation's first project was "Operation Green Turtle." In this effort, thousands of newly hatched green turtles were taken from the protected, transplanted nests at Tortuguero; of these, 20 percent were tagged with plastic disks through the rear flipper and released in order to repopulate the now vacant seagrass pastures of the Caribbean. Carr's assistants, Larry Ogren and Harold Hirth, expanded the small hatchery they had built at the Tortuguero research station and used a Cessna four-seater airplane to deliver hatchlings to various sites around the Caribbean. The Cessna was soon replaced by a US Navy Grumman Albatross seaplane, complete with an eight-man crew under contract with the Office of Naval Research, arranged by Brotherhood member Jim Oliver.

By 1962, 20,000 hatchlings were being sent to dozens of sites where the beach could be protected, in the hope that the long-gone rookeries could be reestablished. Another 8,000 hatchlings were released in Tortuguero. The navy seaplane allowed delivery in the shortest time possible after hatching, thus keeping mortalities to a minimum. The publicity surrounding the deliveries and the international restoration effort prompted fisheries officials in Florida and Mexico to start research programs and created public support for conservation among Caribbean islanders.[15]

FIGURE 11 Children at Tortuguero lagoon, Costa Rica, watching US Navy Grumman seaplane arrive to transport green turtle hatchlings for Operation Green Turtle, 1961.
Special Collections, George A. Smathers Libraries, University of Florida, Gainesville, courtesy of Carr Family Trust.

Carr was very grateful for the publicity Parsons gave to Operation Green Turtle. In the foreword he wrote for Parsons' book, Carr took the opportunity to express his admiration to Parsons' field:

> A geographer is a man to envy. Being by definition a student of the earth, he is free to go anywhere he can get a ticket to and tell of almost anything he can understand. When a geographer goes to a place like the Caribbean island of San Andres, say, and comes away and writes about it, he doesn't have to stick to rocks, geckos, or folk songs as some specialists do. He can report on anything he wants to. One function of geography is to account for man as a feature of the landscape—and what is not grist for such a mill as that?[16]

Carr also let Parsons use as an endplate a map illustrating the initial tag returns, something he was working on for his fourth note on green turtles in the American Museum *Bulletin*.[17]

Disagreement by Proxy

When Parsons' book appeared, the reviews were numerous and very positive. John Hendrickson wrote a review for *Science*. He had just taken up a new position as dean of student exchange at the East-West Center at the University of Hawaii at Manoa. Like Carr's foreword, Hendrickson's review was very complimentary about the field of geography and its potential to contribute to the understanding of a species. He wrote:

This attempt to summarize the totality of man's relationship to an important reptilian species shows to advantage the pleasing range over many disciplines that makes good geography such a valuable contribution to scientific literature. History blends with anthropology and sociology; ecology, zoogeography, and economics integrate with these and other disciplines to tell a fascinating story which preserves scientific accuracy but escapes the monotony of the separate specialized source works drawn upon.[18]

As an active geographer, Parsons must have been pleased with these comments.[19] But he must have scratched his head over Hendrickson's other comments. Hendrickson wrote: "Modern Thais may *say* that the eating of turtle is 'unthinkable' (p. 9), but my Thai friends agree that their countrymen fail by the tens of thousands to *demonstrate* this aversion." Hendrickson also pointed out a mistake in the caption to one of the photographs Tom Harrisson had supplied: "Old Sarawak turtle hands will no doubt be confused by the 'Semah' photo on page 10, which presumably was taken on the nearby coast of the main island of Borneo rather than on Palau Talang Talang Besar." He also noted Parsons' omission from the distribution map of a large and unexploited nesting population at a beach a few miles outside of Karachi, Pakistan, which he himself had visited with some Thai colleagues. He then excused the lapse because the beach was not reported in the scientific literature.

The comment about the Semah photo was clearly a dig at Tom Harrisson, who probably sent the photo and caption to Parsons in his characteristic haste and desire for publicity. Hendrickson had not forgotten how badly Harrisson had treated him. When he saw that Archie Carr had given Parsons a map showing the locations where the first turtles tagged at Tortuguero had been captured, it reminded Hendrickson that

Carr had seemed to question his observations at Talang Talang Besar, perhaps influenced by a disparaging remark from Carr's correspondent, Tom Harrisson.[20]

In his fourth note in the *Bulletin of the American Museum of Natural History*, under the heading "Group Movement and Site Tenacity" Carr had described Hendrickson's finding that at Sarawak the female turtles were "returning for renesting back to the same island (and by inference, to the island where they themselves had hatched)." Carr then wrote: "Although Hendrickson said, unaccountably (p. 503), 'It is not known yet whether the individual female turtle consistently returns to the same island previously used for nesting . . . ,' he had, as we have said, earlier shown strong evidence to the contrary, and we suppose the quoted statement to be a slip of some sort. In another section he points out that, of several thousand observations of females returning to renest in a given season, only 3.7 per cent changed islands and suggested that these might have been frightened away from their preferred island by patrol activity."[21] Carr had then concluded that his data and observations gave little doubt that the green turtle engaged in group travel guided by a strong site tenacity and "homing" capacity.[22]

The quibbles in Hendrickson's book review might also have rankled Carr, as he felt it was as much his book as Parsons'.[23] Carr knew that Hendrickson was still on bad terms with Tom Harrisson and could not therefore get back to Sarawak; but their disagreement had nothing to do with Parsons, and the mistaken caption did not warrant correcting in a publication such as *Science*. Besides, the Parsons book was excellent. But Hendrickson felt strongly that the question of true navigation had not yet been answered. He thought Carr was jumping the gun, using Parsons' publication of *The Green Turtle and Man* to publicize very preliminary evidence. Carr's data, made possible by Hendrickson's flipper tag idea, suggested that the green turtle made purposeful navigations between nesting beaches and the distant feeding pastures of *Thalassia* turtle grasses, as the Cayman turtle captains had told Carr. If Hendrickson indeed thought so, he was correct: Carr was also at work on a general audience book on reptiles for the *Life* magazine library. He used that book and every other publishing opportunity to make the case that the green turtle's large-scale migrations were both a cause for wonderment and a source of peril to the world's "most valuable reptile."[24]

Despite the favorable reviews, Parsons' *Green Turtle and Man* was a wake-up call that went unheeded. A brisk business continued at turtle fisheries and canneries, including those in the Caymans-Nicaragua-Key West triangle whose operations were proving so lethal to Carr's tagged population at Tortuguero. As society's middle class grew, the demand for products like turtle soup grew as well. Soon, turtle soup would be joined by another luxury product made from turtles: women's shoes made of exotic reptile leathers. Carr believed a concerted campaign was needed because Parsons' wake-up call had not had the same emotional impact as another book published in 1962, Rachel Carson's *Silent Spring*. That book raised the specter of extinction, not by human appetites but by the pesticide DDT, an industrial convenience rather than a necessity. People liked the things that harvesting turtle flesh made possible; turtles were food, not poisons on the landscape.

CHAPTER SIX

A Turtle Flap in London

By 1962, Tom Harrisson was convinced that the green turtle population of Sarawak was crashing. It was true that his Turtle Board staff had collected over a million eggs and sold them at one pound sterling per one hundred. But his predecessor, Edward Banks, had recorded an annual harvest upwards of four million eggs. Harrisson's annual egg numbers were declining steadily and at an accelerating rate. To stem the loss, he decided to expand the conservation program on the Turtle Islands. Although the Turtle Board expected him to get the maximum revenue from the eggs, Harrisson nevertheless told his staff to replant a full 10 percent of the eggs collected from the nests into the hatchery, and to do so with even more care. He was delighted later when the hatch rate reached 70 percent. This was far better than what Hendrickson had gotten in 1954. And Harrisson had the staff protect the hand-reared hatchlings for longer than Hendrickson had and take them farther beyond the reef before they released them.[1]

The data resulting from Hendrickson's tagging studies were beginning to alarm Harrisson. They showed that the females nested multiple times within a given nesting season. The nesting population's size was therefore much smaller than Harrisson initially assumed. They showed too that once the nesting season was over, the turtles moved away great distances. Harrisson believed their migrations might be comparable to that of the great whales. If this were so, the turtles could be swimming as far as the treacherous waters of the Pacific around Christmas and Johnston islands, where atomic bombs were being tested. If his sea turtles were to be saved from extinction, they would require protection in the distant waters where they spent the bulk of their time. He wrote, "Alas, there is plenty of reason to think that outside our waters their modern troubles really begin."[2]

It took Harrisson quite a bit longer to realize that the Turtle Board's egg harvesting itself might be reducing the numbers. He had agreed with

Hendrickson's conclusion that egg collection was a potentially sustainable use of the turtles, whereas killing them for their flesh was decidedly not. Harrisson attributed the beginning of the egg decline to the mistreatment of the rookery during the Japanese occupation before he became curator. After that, the turtles' habitat was degraded by the rapid economic development following Rajah Brooke's transfer of sovereignty and the concurrent decline in the authority of the Turtle Board. Harrisson no longer controlled the waters around the islands; they were now teeming with peril for turtles, from motor boats and fishing fleets, to pollution and the bright lights that shone from shore at water level.

To Harrisson's mind, the value of green turtle flesh as food was the main reason for the decline. To combat this peril, he decided to publicize the green turtle's plight to a larger audience than was likely to be reading his beloved *Sarawak Museum Journal*. An active member of Britain's Fauna Preservation Society, Harrisson sent an article to his friends on the editorial board of the society's quarterly, *Oryx*, entitled "The Present and Future of the Green Turtle." He quoted Carr as having called *Chelonia mydas* the "most valuable reptile in the world," but wrote: "Unfortunately, this value is reflected again in the great reptile's alternative common name 'the Edible turtle'. Under the richly ironic scientific name of *mydas*, this ponderous marine has poured its rich juices into centuries of banquets proffered by the Lord Mayor of London or President of the United States. Genuine turtle soup is still a top status symbol in upper-class diet through most of the civilized world—except South-east Asia."[3] Harrisson believed the taboo against eating turtle flesh had saved the green turtle in that region but was, sadly, now beginning to break down because of population pressure and the disturbance of traditional conservation laws and patterns.

Despite his numerous projects as museum curator and government ethnologist, Harrisson's particular passion was wildlife conservation, something he shared with the woman he had married. Barbara Harrisson's specific concern was for the orangutan. She established a sanctuary for orphaned orangutans in the Harrissons' home in Kuching and another in the mountains of Borneo to protect them from the depredations of the pet and zoo trade. Commercial collectors often killed the mothers to get the younger, more easily transported animals, many of whom expired before reaching the zoo.[4] When Tom Harrisson saw advertisements for turtle-skin shoes and handbags, he decided that the time was right to depopularize luxury products made from turtles. This would build support for the international trade restrictions that were going to be needed to prevent species like the green turtle and the orangutan from extinction.

By the 1960s, wildlife conservation was in full swing as a global campaign. Many of the people Tom Harrisson most respected in England were at its center. Meetings of their organizations gave him reason to return often to Europe. There, he would meet former Oxford mentors and patrons in the conservation community and help mobilize action for vanishing wildlife. As an added benefit, he could scout for the new appointment he would soon need when he retired from his civil service post in Borneo.

The British conservation community welcomed the well-known explorer and ethnographer. But Harrisson brought more than celebrity to its cause. He had already worked with many of the principal figures in conservation: Julian Huxley, the eminent biologist; Max Nicholson, Britain's leading ornithologist; and Peter Scott, son of the famous and ill-fated Antarctic explorer Captain Robert F. Scott and in his own right a champion sailor, painter, and pioneer in bird conservation.[5] British conservationists welcomed Harrisson's insight on tactics for mobilizing public support and for the boundless energy and network of connections he brought to the growing international organizations they were building.

Whenever Harrisson was in London, he peppered the media with commentary on nature conservation, taping radio and television spots and sending guest articles to London newspapers. He chided zoos for causing the decline in orangutan populations in Southeast Asia. Julian Huxley, Harrisson's friend from Oxford and then secretary of the London Zoological Society, acknowledged that zoo collecting was indeed part of the problem facing wild species, especially in Africa. When Huxley penned a series of articles for the *Observer*, Nicholson and Scott began discussions which led to the formation of the World Wildlife Fund in 1961, to raise funds for endangered wildlife.[6]

A Meeting in Nairobi

Huxley and Nicholson encouraged Harrisson to get involved in the International Union for the Conservation of Nature and Natural Resources, or IUCN, which Huxley had helped to found when he was UNESCO's first director-general. A hybrid organization whose members were both governments and nongovernmental organizations, IUCN's role, in Huxley's words, was to work to prevent civilization from "destroying the cosmic capital on which we live," especially by preserving species threatened with extinction.[7] IUCN had joined the newly formed World Wildlife Fund to set priorities for funding local conservation interventions around the globe to prevent species extinction.[8] When IUCN held an assembly in 1963 in Nairobi, Harrisson decided to go so that he could lobby on behalf of the wildlife of Southeast Asia. He also hoped to find funds to continue and expand the museum's conservation work in Sarawak.

Archie Carr also attended the Nairobi meeting. He was there to publicize what Costa Rica was doing to protect the green turtles at Tortuguero. Four months before, the government had issued a decree extending the ban on commercial captures of turtles and eggs from the high tide line out to one kilometer from shore. Carr proposed a resolution to the assembly expressing appreciation to Costa Rica for this action and urging other countries to follow Costa Rica's example. Carr hoped international recognition would inspire Costa Rica to declare a national park for the entire length of the beach, from Parsimina to the mouth of the Tortuguero River, where his research station was located. This would make it unnecessary for the Caribbean Conservation Corporation to renew the lease for the three-mile stretch of beach.[9]

When Carr and Harrisson met for the first time at Nairobi, they compared notes on their green turtle populations, resuming a conversation that they had begun in correspondence in the early 1950s. They joined the assembled members in supporting a resolution restructuring IUCN's Survival Service, and Harrisson was particularly happy when the assembly elected his friend and fellow birder Peter Scott as chair of its renamed Survival Service Committee. Scott announced that he would immediately broaden the committee. Both Scott and Max Nicholson, though devoted to birds, believed the Survival Service Committee had to branch out beyond birds and mammals. Scott described his plan to create groups of specialists, chosen on a scientific basis from people who had wide contacts in the field and could collect information on a broad range of species that were facing survival problems. The Survival Service Committee would also include an Alert Group, a small action group that would meet quarterly to respond to urgent matters as they arose.

Harrisson accepted with pleasure when Peter Scott asked him to join the Alert Group, and he endorsed the members' decision to adopt a new classification scheme to help the specialist groups categorize their species. These categories were based on the degree of danger of species' extinction:

1. very rare and believed to be decreasing in numbers;
2. less rare but believed to be threatened, and requiring watching;
3. very rare but believed to be stable or increasing;
4. status inadequately known; survey required or data sought.

After the World Wildlife Fund was created in 1961, Scott had created a system of color-coded notebooks to help it and IUCN's various committees set priorities for action. For species threatened with extinction and classified according to the scheme, the Survival Service Committee would compile a red data book. With information from the specialist groups, the committee would use asterisks to convey the degree of risk: one asterisk meant that steps toward better conservation were required "when possible"; two asterisks, that steps were required "rather urgently"; and three asterisks, that steps were required "very urgently."[10]

To address the wild animal trade, the Nairobi assembly called for the development of an international treaty on trade that would control the collection of wild animals for the pet industry, for zoological parks, for laboratory research, and so on. Tom and Barbara Harrisson had stressed that the prices zoos were willing to pay was leading to the extinction of the orangutan.

When the Nairobi meeting was over, Scott wrote to Archie Carr, whom he had met there, asking him to chair a specialist group on marine turtles. The group's job would be to classify the extant species and populations according to the Red Data Book categories.[11] Carr accepted with some trepidation. The task would give him a chance to reiterate the dangers these species faced. But the job might require dealing with the uncertainties in the taxonomy of the sea turtles, something the new specialist group was probably not able to tackle, meaning the job would fall to him and his students.

Tom and Barbara Harrisson kept close tabs on the drafting of the treaty; when the first draft was circulated for comment, Barbara analyzed it on behalf of the primate specialist group and kept that group informed of the treaty's progress as the drafting process dragged on for several years.[12] She believed the orangutans could not wait for the treaty to be adopted. She used her network of primate specialists to get a resolution introduced at the annual meeting of the American Association of Zoological Parks and Aquariums in Mexico City in March 1967. She convinced the members to get ahead of the international requirements and adopt their own self-policing scheme. The scheme would involve a list of wild species that the members agreed to never buy or accept as gifts because they were endangered or were about to become so.[13]

From Gourmand to Glutton

Although he too had supported the treaty idea at Nairobi, Tom Harrisson believed it was necessary to change public tastes and attitudes before there would be any lessening in international trade in wild animals. He thought it was naïve to think international conclaves could save species: "It is mere pious hope, wishful-thinking indeed, to believe that international symposia can cure this sort of ill. Nor will lip-service to legislation by governments achieve anything, except international self-satisfaction." One had to act at the local level and not use the fact that outside actions were at work as "an alibi for local neglect."[14]

In the 1930s, Harrisson had co-founded Mass Observation, the first large-scale survey of British public opinion. Now, using methods from Mass Observation, he decided to put his principle of "local action" into practice and reduce the popularity of turtle soup through a media campaign of letters to the editor, interviews, and articles. Consumption of turtle flesh, whether in soups, shoes, or ladies' handbags, would become a stigma instead of status symbol.

He had actually begun this campaign before the Nairobi meeting after reading an article in the London newspaper the *Observer* describing chefs cruelly butchering a green turtle for the soup kettle. He immediately wrote a stinging letter to the editor, who happened to be his old friend David Astor, enclosing a copy of his recent article in *Oryx*. He warned the newspaper that it had probably given offense to its readers in those regions of Asia east of India, "where turtles are sacred." Furthermore, this senseless slaughter would lead to the turtle's extinction.

Astor wrote back and invited Harrisson to write a full-length article on the subject. Harrisson put him off, noting that for the foreseeable future he would be too busy dealing with the local impacts of the political confrontation between Malaysia and Indonesia. People in the highlands for whom he felt responsible were at risk from these troubles, and he felt obliged to resume the role he had played in World War II, that of a guerilla leader and "jungle wallah."[15]

Although Harrisson's writing lacked some of the charm of Archie Carr's, it could hit its mark just as accurately. When Harrisson finally wrote the full-length article, "Must the Turtle Die?" it was published, not in the *Observer*, but in both the *Sunday Times of London* and the German magazine *Die Umschau* in June 1964. In it, he described the conservation project at the Turtle Islands, the Semah turtle fertility celebrations, the green turtle's nesting habits and the near veneration of turtles in Asia, save for their eggs. And he sounded the alarm: the taste for turtle soup had passed from the gourmet to the greedy, as a symbol of the affluent society: "One Frankfurt firm sells nearly a million tins a year, and you can get 'genuine' turtle soup for two marks in remote pubs high in the Black Forest." But there was no truth to the matter that turtle soup was a tradition of long standing. He could find no record earlier than 1850 of Guildhall banquets with turtle soup.[16]

The leading maker of turtle soup, LaCroix, of Frankfurt am Main, Germany, wrote a letter to the editor of the *Sunday Times*, objecting strenuously to Harrisson's insinuation that the soup industry was causing a species' extinction. The company also wrote directly to IUCN headquarters in Morges, Switzerland, to ask if Harrisson's was the official position of the organization. Did the organization stand behind the factual claims of the prominent member of its Alert Group?[17]

Turtles in Trouble in Paradise

Meanwhile, a taste for turtles was being promoted elsewhere in the West, far from the damp, gray weather of London and Frankfurt. Green turtle steak was becoming a popular dinner item in the new hotels and restaurants along Waikiki Beach, in Honolulu. Statehood had brought Hawaii a boom in construction of hotels for the new tourism industry made possible by jet planes and the growing affluence of the postwar United States.[18]

Honolulu was also becoming a center for research and education on the peaceful cooperation between the West and the East. In 1963, shortly before Malaya became part of the federation of Malaysia, John Hendrickson left his university post in Kuala Lumpur. He moved his family to Honolulu and took up a position as vice chancellor for student interchange at the East-West Center, a position he had hoped would open up for several years. Although world peace through educational exchange was his new mission, he worked as a voluntary curator for the herpetological collection at the Bishop Museum. He published papers from his research on the sea turtle nesting beaches of eastern peninsular Malaya in the *Bulletin of the Raffles Museum* (Singapore). He was especially interested in the nesting behavior of sea turtle species whose eggs were heavily exploited in Malaya under a system of exclusive rights that the government awarded to the highest bidder. Before leaving Malaya, as part of these studies Hendrickson had started a hatchery at a leatherback nesting beach.[19]

After four years of academic administration at the University of Hawaii, Hendrickson was ready to go back into research. He was especially intrigued by mariculture, the farming of marine life, and its potential for alleviating hunger and reducing pressure on wild fishes and marine life. He took a job as the director of the new Oceanic Institute at Makapuu, a few miles from lush Manoa Valley, where the East-West Center and the University of Hawaii campus are located. At the Oceanic Institute, Hendrickson began studies of the nutritional requirements and diseases of candidate culture species, including the green turtle, and also studied shark migration.[20]

Hendrickson's interest in turtle mariculture led him to undertake a survey for the South Pacific Fisheries Program and the UN's Food and

Agriculture Organization. Joining him on this survey was Harold Hirth, Carr's former student, who had assisted with the early green turtle research at Tortuguero and Ascension Island. Hirth and Hendrickson were looking for islands where the conditions were favorable for turtle farming. In Fiji, they met a young Peace Corps volunteer named David Owens working on fisheries development who was particularly interested in the possibilities that turtle culture presented.

Green turtles had been commercially exploited in the Hawaiian archipelago ever since the mid-nineteenth century, when advances in navigation aids and better seamanship made possible purposeful trips to the coral atolls and tiny islets that lie leeward of the main islands. Less successful navigators who had shipwrecked their crews were more than happy to feast on the easily caught turtles that basked on the islands.[21] The Hawaiian people had long caught turtles by turning them on the beach or diving on them from canoes and kept them in fish ponds. The Hawaiians had learned to kill turtles with harpoons from the whaling ship crews, especially the Tahitians from Bora Bora, who had given up whaling and settled on the island of Maui.[22]

The northwestern Hawaiian Islands, where the green turtles nested, were now part of the Hawaiian Islands National Wildlife Refuge and under the administration of the US Fish and Wildlife Service. At the request of FWS, Hendrickson took a look at the tagging data from 1964 to try to estimate the population size. Although there were serious limitations in the data, Hendrickson estimated that the colony contained 2,600 to 5,000 turtles and that the population was a fraction of what it had been a hundred years before.[23]

Hendrickson was acquainted with Hawaii's governor, John Burns. As a member of the US Congress, Burns had supported establishment of the East-West Center and was a friend to the center while Hendrickson was its academic dean. So when an article appeared in the local newspaper on how much wildlife hunting and fishing had grown in the state since the end of the plantation economy, Hendrickson decided to write the governor. He complained that the state of Hawaii was showing that it "could not care less" about how much wildlife was being killed. He noted that more was being done to control turtle killing in the distant former colonies of Sarawak and Ceylon than in Hawaii. Even Florida was doing more for its turtles. Hawaii's once-sizable population of green turtles was probably now closer to extinction than any other turtle population in the world.[24]

Thus, when Archie Carr invited Hendrickson to join IUCN's new Marine Turtle Specialist Group, Hendrickson welcomed the chance to contribute what he knew of turtle populations in Hawaii as well as in Malaysia. It would not matter if Tom Harrisson, as co-chair of the Specialist Group, carried on as if he knew everything about the green turtles of the South China Sea. Hendrickson now had a different sea turtle population to study and to fight to conserve.

The Buffalo of the Sea

Publication of his own book, *The Reptiles,* in the Time-Life Nature Library in 1963 gave Archie Carr an occasion to sound the alarm about the disappearing green turtle. It also was a chance to raise, if not answer, a question he would later cross swords over with other turtle scientists, especially John Hendrickson. The question was not about the turtle's biology and long-term migration but about the best means to promote the edible turtle's conservation. Given its value as a food resource and the growth of coastal human populations, could farming the green turtle prevent its extinction? Carr wrote:

Another field in which the information provided by the green turtle investigations [at Tortuguero] will be useful is turtle farming. The submarine plants the green turtle feeds on grow in pure stands in shallow water behind reefs, or on shelves among islands. Some of these places lend themselves to being easily fenced. In such enclosed natural pastures, green turtles could be kept like aquatic cattle. The Caribbean Conservation Corporation is planning pilot projects of this kind. If they are successful, the green turtle may become one of the first marine vertebrates to be successfully cultured for food.

But Carr was very ambivalent about relying on utilitarian reasons for conserving species:

When I hear of a new idea for raising the food yield of the sea—or of the land for that matter—I have mixed feelings. I hate to see any comfort come to those who encourage the useless multiplication of man. But there are very hungry people in places where green turtles once were abundant and are now unknown. It is the sea that will be called on to feed these people, as well as the hordes of our descendants. And anyway, when you save a species for meat you are bound to save useless bits of wilderness to go along with it. So whenever a bit of the waning world can be saved for meat, then go ahead and do it that way. But inadvertent savings of scraps will never keep off the ruin of the earth. The only way is to name the real obligation clearly, to say without hedging that no price can be set for the things that have to be preserved. When New Zealand set up the *Sphenodon* preserve I wish they had said it was not to placate zoologists, but so that plain men could go on singing the tuatara out of its hole.[1]

Archie Carr had thought turtle farming was a worthwhile conservation strategy throughout the 1950s. He believed that once the green turtle's reproductive biology was known, it could be used to encourage the green turtle "to regain something approaching its primitive range and abundance."[2] He made a similar comment in his address to the American Institute of Biological Sciences when it met in Gainesville, and he included the idea in *The Windward Road*.[3]

By 1957, Carr was thinking about putting some form of farming into practice. That spring, David Caldwell gave a paper on their study of turtle fisheries of Florida at the North American Wildlife Conference. Landings of green turtles in Florida consisted of mostly large adults caught in Costa Rica and Nicaragua. Because Florida no longer had a breeding population of its own, only a very small green turtle commercial fishery persisted on the west coast of Florida, based on a seasonal population of juvenile greens born somewhere in the Caribbean. Clearly, what was needed was a complete ban on turtling in the United States.

Caldwell read with conviction Carr's forthright conclusion: "There is no reason, except political expediency, half-hearted action, or lack of foresight, why anyone should be allowed to kill a nesting turtle or rob a nest anywhere on the shores of the United States. It is a poor way to make a living, and in fact no real living is made that way. This activity has not only lost us a once valuable and potentially even more valuable resource, but it must necessarily prevent any natural resurgence of that resource."[4]

Caldwell explained that despite Florida's law protecting nesting females while they were on the beach, poaching was common, as enforcement resources were spread thinly across the miles of unpatrolled beaches. And a loophole allowed female turtles to be taken while they were still in the water, even before they had emerged to nest. Archie Carr had always taken a dim view of the way in which Florida had neglected its green turtle fisheries and said as much in this paper and in several other books and papers. As far as he was concerned, the neglect started in 1896 when the state had completely ignored the US Fish Commission's warnings.

Greater enforcement and "providing sanctuaries for reproduction" would not be enough, however, to restore Florida's once plentiful breeding population of green turtles. Caldwell predicted that eventually an active management program of "restocking, or even farming," would be done. While the eggs are easily damaged in transport, the hatchlings are not and could be easily shipped from a productive nesting beach to places like South Florida, held in enclosures and raised to a weight of a pound or two, and then released onto the ungrazed "fallow turtle pasturage. While such a project involves too many uncertainties to be justifiable on a big scale, it is certainly worth trying experimentally, and we venture to suggest that it is very likely the model for important moves to come."[5]

When Caldwell finished his talk, Harold Coolidge from the US National Research Council asked the lead-off question. Had anyone thought of farming turtles on a commercial scale as part of the Florida fishery? Not to his knowledge, Caldwell replied. The closest example he was aware of was in the East Indies where, owing to religious beliefs, the people only ate turtle eggs. The government had a hatchery at a beach where eggs were collected, reburying some proportion of eggs that were dug up, letting them hatch, and then releasing the hatchlings offshore once they reached one or two pounds in weight.[6] Caldwell was speaking of the Turtle Islands of Sarawak.

Caldwell's response prompted Coolidge, who was heavily involved in wildlife conservation and had founded the Survival Service Committee (later Commission) of the International Union for the Conservation of Nature and Natural Resources (IUCN), to remark on the conservation implications of the turtle research John Hendrickson of the University of Malaya had been doing in the islands off Sarawak. Not realizing that Caldwell was also referring to the work at Sarawak, Coolidge mentioned that Hendrickson estimated that if green turtle hatchlings were reared in captivity for three months before releasing them, their survivorship increased from about 1 percent to roughly 60–70 percent. Coolidge said he understood, however, that the hatchery at Sarawak had yet to adopt this period of captive rearing despite the importance to Sarawak of the revenue from the turtle eggs. He also commented that he was in discussions with the South Pacific Commission of an international turtle conservation program based on Hendrickson's recommendations, and that he hoped similar programs might be developed in the Caribbean region and in Florida.[7]

The vision Coolidge described was soon to be realized when Carr and the Caribbean Conservation Corporation began Operation Green Turtle in 1959. Localities that received hatchlings from the hatchery at Tortuguero included several in Florida.

The Burden of Proof

While experimenting with restocking strategies, Carr continued to assemble the "strong evidence" that green turtles engaged in purposeful, open-ocean migration of the kind that Hendrickson had not seen in his study of the South China Sea green turtles. Carr found the perfect place to obtain the proof when he learned of green turtles feeding in the coastal waters off Brazil. There was no obvious nesting rookery along the Atlantic coast of South America, but there was a nesting colony on Ascension Island, an isolated volcanic island due east of Brazil lying midway between South America and Africa. He sent his research assistant Larry Ogren to Ascension to set up a tagging study to supplement the tagging work Ogren had done for Carr at Tortuguero. Because there are no landmarks between Brazil and Ascension and the prevailing current flows in a westerly direction, Carr reasoned that any commute between the two points had to be by purposeful navigation.

When turtles Ogren tagged on Ascension were captured 1,400 miles away grazing off Brazil, and others returned to the same tiny beach coves on Ascension where they had been tagged, Carr was convinced; he knew, however, the evidence was still circumstantial. "This does not prove the reality of the migratory pattern beyond all possible doubt, but I think it does so beyond a reasonable doubt."[8] To remove all doubt he needed to explain how the turtles did it. Carr needed two more breakthroughs: a way to track individuals to determine what course they followed and a method to test turtles' senses to find the specific mechanism responsible for their marvelous feats of navigation.

With Larry Ogren's help, Carr had shown that hatchlings used a complex chain of sensory cues and responses to emerge from the nest and find the sea.[9] What he needed now was an experimental approach that could incorporate ideas from cognitive psychology and sensory biology to whittle down the list of possible mechanisms involved in the turtles' navigation.[10] His prospects for getting such help brightened considerably when his graduate student David Ehrenfeld brought to the field station an experimental approach based on his former medical studies. While Ehrenfeld tested the turtles' vision and color reception, Carr and his other students focused on tracking turtles' direction and speed when they left the nesting beach. With these projects, Carr was confident he could make progress on these questions.[11] He was particularly optimistic when a talented, young experimental psychologist, Nicholas Mrosovsky, joined them at Tortuguero in October 1965. Ehrenfeld invited Mrosovsky to use Ehrenfeld's research staging areas, and Mrosovsky devised his own as well to test the role of light in the sea-finding behavior of hatchling turtles.[12]

The tagging studies and hatchery work at Tortuguero were labor intensive, but for lots of people it was a labor of love. Carr's grant money supported only a limited number of graduate research assistants, and the Caribbean Conservation Corporation could afford only one full-time staff member, Leo Martinez. Nevertheless, the opportunity to work with Carr had so much appeal that Carr had little difficulty recruiting students from the University of Florida and other colleges and universities.[13]

The key was to keep the research station operating long enough to make progress on these questions. This would require both public support and the willingness of the political leaders of Costa Rica to maintain and extend the laws protecting the nesting beach at Tortuguero. Carr believed the best way to keep both private and public support was through a steady campaign of good publicity to continuously stoke the public fascination with the animal world and the amazing story of the green turtle. Geographer James Parsons had begun this process with his photographic essay on the work at Tortuguero and the effort to restore former rookeries in the Caribbean. Operation Green Turtle was itself some of the best publicity possible. There was always a happy crowd of onlookers whenever the navy men in their orange jumpsuits delivered a new batch of baby turtles. Onlookers attracted newspaper photographers, and Carr still had the inclination to use his writing gifts to keep the green turtle in the public eye.[14]

A Debate Begins

Archie Carr knew that one of the best venues for publicizing vanishing elements of the natural world was the popular magazine *National Geographic*. Carr wrote a compelling article for the magazine on the plight of the American alligator,[15] and its editors encouraged him to write another on the green turtle of the Caribbean. To get the best photographs, Carr invited fellow Floridian and marine biologist Bob Schroeder to visit the research station at Tortuguero. An accomplished underwater photographer, Schroeder had himself recently published an article in the *National Geographic* on coral reefs, as well as a book on undersea life.[16] And although he had never been to Tortuguero, Bob Schroeder already knew lots about green turtles.

In the early 1960s, a student who had worked at Tortuguero asked Carr if he could take a dozen hatchlings back to the University of Miami, where he was doing graduate work. When the student's studies took a different direction, he gave the baby turtles to Schroeder, who was just finishing his own PhD research raising gray snappers to study their parasites. With the help of Jean Schroeder, his wife and partner, who shared his knack for raising marine organisms, Schroeder set out to raise the turtles in the boxes he had still floating off the dock at his home in the Florida Keys.[17]

The Schroeders soon had another full-scale research project under way. When the baby turtles refused to eat turtle grass (*Thalassia*), the Schroeders realized that green turtles began life as tiny carnivores and offered them a variety of fish scraps and other foods, including canned dog food, all the while monitoring their health and measuring their growth rate. After a year, the turtles had grown to the size of dinner plates. Because so little was known about the lives green turtles led after the hatchling stage, Schroeder began corresponding with Archie Carr, who encouraged his studies and offered to send him more hatchlings from Operation Green Turtle. Schroeder tested the young turtles' reaction to sargassum, the seaweed that forms large offshore mats on which Carr thought hatchlings might spend their "lost year." Impressed with the little turtles' hardiness and the relative ease of raising them, Schroeder began to think seriously about farming green turtles.[18] He knew green turtles were still being fished in Florida and killed by poachers for their meat and calipee elsewhere in the Caribbean. If farm-grown turtle products could be sold for a modest price and undercut the poachers, it might just relieve this drain on wild populations.

Schroeder mentioned his idea to Carr and was eager to continue the conversation as he learned more about raising green turtles. He helped Operation Green Turtle by taking under his care 2,000 hatchlings stranded at the Miami airport when the navy's seaplane developed engine trouble. Although he and his wife had recently moved to larger quarters in Lower Matecumbe Key, he was running out of space. He contacted Carr about a suitable release site for the young turtles, and they settled on Cape Sable beach in the Everglades National Park. It was remote and protected, yet close enough for local news reporters to witness the release and further publicize the operation. The Carrs and the Schroeders finally met in early 1966 when the Carr family helped take the turtles to their release at Cape Sable.[19]

A few months later, Carr invited the Schroeders to do the photo shoot at Tortuguero. Like Carr, Bob Schroeder found great satisfaction in sharing the wonders of the marine world with the general public, so he readily accepted. Schroeder had recently published an article entitled "Buffalo of the Sea" for a popular magazine on ocean science, which made the case for raising green turtles for food.[20] The trip would be a chance for the Schroeders to learn more about Carr's research on migration and for them to continue the conversation they had started about turtle farming.

The research was indeed fascinating and made for wonderful photographs. David Ehrenfeld was experimenting with specially fitted goggles on the large females, testing whether they found the sea after nesting by following polarized light or light of particular wavelengths from the sky over the sea, or the brightness of the horizon. When one goggle-fitted female sprayed Ehrenfeld with sand as he tried to turn her right-side up, Schroeder got an excellent photograph.[21]

The people assembled at the research station discussed many things about the conservation challenges surrounding the green turtle, including the arguments for and against farming as a way to end turtle hunting. Costa Rica had recently prohibited taking turtles or eggs from anywhere on the entire length of the beach, but a loophole allowed harpooning of turtles beyond the line of breaking waves.[22] The demand for green turtle soup was growing; and a skilled poacher could remove the calipee from under the plastron of a turtle, leave her dying in the surf in the dark of night, and run almost no risk of detection. And marketers were developing demand for new products. Poachers were being offered a good price for turtle skin by dealers in exotic skins and leather, who marketed the handbags as the latest fashion and status symbol. A cosmetics company in California was marketing green turtle oil as a beauty aid to the wealthy, a new product that Carr found particularly irksome.[23]

Carr and Ehrenfeld were both ardent conservationists and were worried about the impact the burgeoning human population was having on natural places and species. They talked into the night with the Schroeders, debating whether farming could put a dent in this drain on wild populations. If farming was technically feasible, could it be done cheaply enough so that the farm could sell green turtle skin, fat, and calipee for less than the poachers were being paid?[24]

Carr and Ehrenfeld already knew quite a bit about the challenges of turtle farming. Especially the hurricanes. Three years earlier, Carr had sent Ehrenfeld, with 3,000 hatchlings from Operation Green Turtle, to a shallow embayment called Union Creek, on the northern shore of Great Inagua, the southernmost island in the Bahamas. His job had been to set up an experimental culture project on the site of an old turtle farm for the Caribbean Conservation Corporation.[25] Here, Carr could test his idea that turtles could be reared, not in pens or tanks, but in a natural area where *Thalassia* grew in abundance, separated from the ocean by a seawall. With the help of local nature wardens Samuel and Jimmy Nixon, Ehrenfeld had installed a thousand feet of rock wall, blocking sea passage from a four-square-mile area of shallows where juvenile green turtles foraged on the seagrass beds. It became a race against the clock, however, as a tropical depression over the Atlantic Ocean developed into Hurricane Flora. Other severe storms had followed for the next two hurricane seasons.[26]

Ehrenfeld told the Schroeders that when he returned in 1965 with another batch of hatchlings, he was amazed to find 350 green turtles still living within the enclosure. Hundreds of turtles had been washed over the wall, but many were still milling around, grazing on the creek's turtle grass flats. Carr considered it a success, but Ehrenfeld had his doubts.[27]

Long before, somebody else had thought Union Creek was a good place to grow turtles, and the Caribbean Conservation Corporation's pilot farm had been built on its ruins. Carr complained that the former tenant had left no records, so no one knew exactly why it had failed. He had several hunches, however, and hurricanes were only part of the problem. Baby turtles would crawl out of the enclosure onto the beach and either dry up or be eaten. If they survived to the grass-grazing stage, the farm had to find just the right balance between the amount of vegetation available and the number of large turtles allowed to graze it. The water had to be clean and warm and the food supply constant. If the turtles were fed anything to supplement the turtle grass, the supplemental food would foul the water, and the water would then become unhealthful unless cleaned. And predators would be able to get in and out of the area and constantly threaten the livestock.

Schroeder offered his experience on Lower Matacumbe Key as a contrast. The problems of natural enclosures were manageable. By raising the young turtles in his system of floating boxes, he could keep them in water that was the right depth for them to feed and avoid losses at that stage. He could remove the boxes when bad weather or predators threatened. If economies of scale and an adaptive approach were used, he was convinced that turtle farming could be economically viable, even profitable after awhile.

Ehrenfeld thought it would never fly; no farm could ever raise turtles at a lower cost than nature does. Enclosed areas would not be adequately flushed and would quickly run out of pasturage, necessitating raising grass elsewhere and delivering it to the herd. There would still be poaching as long as people paid money for things made from turtles. Unless the farm had a breeding herd, eggs would have to be taken from the existing rookeries. Tortuguero's hatchery was supporting the current experiments in farming and restocking, but the supply was far from infinite, and nesting numbers varied from year to year. By restocking or releasing hand-reared turtles, Operation Green Turtle might one day rebuild the nesting population, but no one yet had evidence that yearlings would have either the instinct or the experience necessary to return to their natal beach and complete the life cycle. Making turtle farming a commercial success made the "labors of Hercules seem easy in comparison."[28]

Carr agreed that if farms proliferated in the Caribbean by growing turtles from eggs collected from natural nesting beaches, the wild rookeries could not stand the drain. Farms would have to develop captive breeding herds. But would a farm stock of turtles even breed in captivity without undergoing the migration? He also questioned whether replacing eggs taken from a natural rookery with pen-reared yearlings might not be "just a laborious way to kill them," as they would not have imprinted on their natal beach and would not know where to go to find seagrass pastures. It was not clear from the research at Sarawak that it was possible to compensate for egg harvests with hatcheries.

Bob Schroeder thought it was apparent that no conventional conservation methods would do much good in Central America. There was too much poverty, the coastlines were too remote, and law enforcement was nonexistent. Everyone expected the number of nesting females to continue to drop each year. Something unconventional was necessary to save the green turtle, which was still, as Archie Carr said in 1954, just "too good to last." Schroeder's idea was to use the knowledge gained from the past twenty years of research—in Sarawak, Malaya, Tortuguero, and Florida—to make another go at farming. He was convinced it was worth a try.[29]

Farming a New Fleet

When Carr's article appeared in *National Geographic* in mid-1967, it did indeed reach a broad audience.[30] The article emphasized the research under way at Tortuguero on the green turtle. But it was the opening sentences that probably had the greatest consequences, directly for Bob Schroeder and indirectly for Archie Carr. A young man who read the article in England was particularly taken by Carr's statement that the green turtle was the most valuable reptile in the world and "offers an expansible food resource for the future. . . . There is a ready market for frozen turtle meat, a growing demand for clear green turtle soup, and a rising commerce in turtle hides for leather."[31] The young man's father, Antony Fisher, had recently sold a successful business, Buxted Chicken

Company, to a multinational food company and was looking for another investment. When son Mark told him of the article, Fisher was intrigued. He could domesticate another food animal, this time an animal that swam in the warm seas where he and his son loved to snorkel and dive. Fisher decided to write to the article's author, Archie Carr. Rather than replying directly, Carr sent the letter on to Bob Schroeder, who had been fielding enthusiastic letters from would-be farmers ever since his "Buffalo of the Sea" article had appeared.[32]

Antony Fisher was more than a business entrepreneur. He was a passionate advocate of the free market and its superiority over state ownership of the means of production in creating jobs. In 1954, he had established the Institute of Economic Affairs in London, the first conservative think tank, founded to promote the ideas of the Austrian free-market economist F. A. Hayek. Flush with the success of their chicken breeding and production business, the Fishers were intrigued by the prospect and challenge of creating a green turtle industry. Antony Fisher had made a fortune by breeding a fast-growing chicken that was affordable to the average household.[33] By 1968, when he sold his company, Buxted was producing 800,000 broilers per week.

But the Fishers did not have the scientific background to understand that there was more unknown than known about the green turtle and did not see the warning signs in what Carr had written. He noted that "two important uncertainties trouble" the restoration program, Operation Green Turtle: "One is that nobody knows how long it takes green turtles to reach maturity. They have not been observed under natural conditions long enough for them to grow from hatchlings to breeding adults. The other uncertainty is whether the transported hatchlings, when mature, have the urge to go back to where we released them."[34]

In July 1968, the turtle pens at the Schroeders' home in the Florida Keys caught the attention of Irvin Naylor, a businessman from Baltimore who was visiting the Schroeders' neighbor. When Naylor asked about the research, Bob Schroeder explained his idea of raising green turtles for food and other products. He mentioned the difficulty he was having in finding financing; he had temporarily shelved his plans for a pilot project on Grand Cayman when his initial investor backed out. Things were not all bad, however. He was hoping to follow up on a letter Archie Carr had gotten from a businessman in England who was interested in turtle farming. And the next day, the Schroeders were expecting a visit from the Cayman Islands secretary of agriculture. They planned to ask the Cayman Islands government for financing and an exclusive license to ranch green turtles in the seagrass-rich North Sound.[35]

As Schroeder spoke, Irvin Naylor realized he was hooked. An entrepreneur accustomed to taking risks, he owned a thriving business manufacturing wooden boxes for cigars and men's aftershave lotion and had built a ski resort with the profits. He asked if he could come back the next day and meet the secretary.

Three months later, in October 1968, the Schroeders' company, Mariculture, Ltd., was incorporated in the Cayman Islands with a capital investment of $10,000 dollars each from Naylor and from his business associate Henry Hamlin, and with the promise of additional funds from Antony Fisher. And the first shipment of turtle eggs was incubating at the Schroeders' rented home on Grand Cayman. Schroeder had gotten a concession from the government of Costa Rica to take eggs from Parsimina near Tortuguero on condition that Mariculture return 1 percent of the hatchlings after a year of hand-rearing and release them on the beach where their nests had been.[36]

Schroeder's design for the farm was based on the ranching idea he described in "Buffalo of the Sea." The eggs would be collected at rookery beaches, transported to an incubation facility at the farm, and the hatchlings hand-raised for one year, until each weighed ten pounds. At that size, they could be fitted with a numbered tag and then released on the seagrass beds to feed on their own and grow to a marketable size of 100 pounds in three to four years. The marked tags would designate the turtles as the private property of the farm. A control group would live in a fenced seagrass pasture, and exceptional turtles would be kept as breeding stock to eventually replace the wild sources of eggs. The farm would build a plant for processing the turtles into products, including turtle meat, soup, oil, leather, and other things that would be developed. If the breeding stock's egg production could be increased, the farm would consider selling turtle eggs.

Schroeder estimated that "with a reasonable amount of advertising, a million green turtles per year could be marketed almost at once." Because turtles lay so many eggs, he estimated that his company could be producing over a million turtles per year within three years. A million to

20 million turtles could be produced by the Caribbean region if the hatchlings were protected; if Mariculture expanded throughout the Caribbean, 100 million turtles a year was not impossible, to the enormous economic benefit of the country that hosted the enterprise. He estimated that the Cayman Islands site was capable of a maximum of 40,000 or 50,000 turtles per year.[37]

Bob Schroeder's ultimate vision for Mariculture, Ltd., was to build facilities in Costa Rica and Nicaragua. He would either close down the operation on Grand Cayman Island or keep it as an experimental facility because his investors and board members enjoyed meeting there. In pursuit of his plans, he traveled to Managua to meet with Manuel Sequeira, president of Nicaragua's institute of national development, known as INFONAC. Schroeder told Sequeira that any country that hosted a turtle farm was sure to benefit greatly. Not only would it save the green turtle from its imminent extinction, but it would also create jobs and a lucrative export commodity.

His company, however, would need Nicaragua to give it two legal protections. First, it would need a license, preferably exclusive, to use Nicaragua's seagrass beds as rangeland for its turtle livestock. In exchange for the license, Mariculture would pay a royalty on the sale of the turtles. Second, it would need Nicaragua to ban the conventional forms of turtle hunting in Nicaragua's waters. Turtle hunters would soon be finding more farm-tagged turtles than wild turtles and, unless forbidden, would kill the company's livestock. Any wild turtles still living on the seagrass pastures had to be spared because they would provide the eggs for the company's initial operations, for which Mariculture planned to pay Costa Rica royalties in exchange for its protection of the wild breeding stock.

This second condition seemed particularly reasonable to Schroeder. Nicaragua had recently rescinded the Caymanian turtle schooners' fishing licenses, after tiring of trying to collect duties on the turtles from the schooner captains.[38] This action would dry up the annual shipment of 5,000 green turtles from Nicaragua to the canneries in the United States and Costa Rica.

Sequeria asked Schroeder to send him more information so he could think about it. He knew little about turtles but agreed they were a valuable resource to Nicaragua and should be exploited. His government was encouraging the construction of turtle processing plants on the Miskito coast now that the Caymanian turtle schooners had been ejected. His agency was financing a turtle processing factory in Puerto Cabezas that would sell turtle meat and calipee to the canneries and skin to the leather trade.

Schroeder and Robert W. Ellis, the regional representative of the UN Food and Agriculture Organization, who accompanied him to the meeting, told Sequeria that they could guarantee that the Puerto Cabezas processing plant would lose money and was a terrible investment. Not only was the green turtle in danger of extinction; soup factories around the world were closing down, and companies were losing their investments. Just the previous year, soup factories in Australia had had to shut down when the Queensland government banned catching turtles for commerce.[39] Only 1,300 female turtles had laid nests at Tortuguero in 1968, down from the 8,000 that had nested in 1965. If the Puerto Cabezas factory processed thousands of turtles year-round as planned, the green turtle would be extinct in a few short years, and there would be no more raw materials for the factory.[40]

By the end of 1969, Schroeder's first shipment of eggs from Costa Rica had been successfully incubated, and the hatchlings were thriving on a dry turtle chow.[41] The Cayman Islands government had adopted a turtle law granting an exclusive franchise to Mariculture, drafted by one of the company's silent partners. In Nicaragua, the turtle processing plant in Puerto Cabezas started up and was soon joined by a second plant in Bluefields. Together they butchered and processed three and four times the number of turtles the Cayman Islands schooners had once taken.[42] If the farm on Grand Cayman was going to supplant this production, it was going to have a hard time catching up. That is, unless something or someone could convince Nicaragua to shutter the slaughterhouses.

CHAPTER EIGHT

Who Will Kill the Last Turtle?

B arney Nietschmann knew nothing of the green turtles in Costa Rica or turtle slaughterhouses in Nicaragua when the spring 1967 semester ended at the University of Wisconsin. He was about to begin a PhD program in cultural geography and was looking for a place to do his field research. During his master's degree studies he had taken a field course with a visiting geographer of Latin America from Louisiana State University named Robert West. The course allowed Nietschmann to live for several months on an old hacienda estate northeast of Mexico City and to draw upon history and anthropology as well as geography to understand the changes in land tenure that preceded the Mexican revolution. For his next research project, Nietschmann wanted to take the same interdisciplinary approach but spend even more time in the field.

Bob West had given Nietschmann a couple of ideas. In 1941, West and fellow Berkeley graduate student James Parsons had traveled the old colonial road through the Sierra Madres that connected Mexico's mining villages to the Pacific coast.[1] Nietschmann wanted to find a similar route where he could retrace the steps of long-ago travelers or traders and determine how global economic forces had changed the physical and cultural geography of Latin America. He was hoping to map out his research trip during the summer, and he was particularly keen to find something that would take him both along the sea and into it.[2]

West told him to consider places in the western Caribbean that Parsons had studied. There were the pine savannas along the coast of Nicaragua where foreign-owned timber companies had hired local people to log the area heavily before moving on to other forests in Central America. Or he could go to the English-speaking islands of San Andres and Providencia, where Parsons had begun to study the islands' connections to natural resource industries on the coast of Nicaragua. But Parsons had not continued his studies of the pine forests. While he was on San

Andres, Parsons had been impressed by the remarkable role the green turtle had played in the islands' economy, and he instead wrote a history of international trade in that species. After *The Green Turtle and Man* was published, Parsons had not returned to the western Caribbean but had moved to other studies and was now chairman of the Department of Geography at the University of California at Berkeley.

As a native of southern California, Nietschmann had been interested in sea life for as long as he could remember. But when he first picked up Parsons' book, he was not sure what a cultural geographer could usefully say about a sea creature like the green turtle. Nietschmann was delighted when he found in Parsons' book a map of the journey he was seeking. Parsons had written of the fishermen of the Cayman Islands—the champion turtlers of the Caribbean—that "there can be few finer examples of cultural conservatism and persistence in the New World than that of this isolated island community of seamen."[3] The maps in Parsons' book showed two things: first, how many green turtle nesting populations around the world had been wiped out by overexploitation; and second, that the nesting turtles Archie Carr had tagged at Tortuguero were being caught by fishermen on the Miskito Bank, a large, shallow area off the coast of Nicaragua covered with dense beds of sea grass and dotted with coral rocks and cays.[4]

As soon as he finished Parsons' book, Nietschmann went to the library to find out more about the Miskito Coast. He found *Waikna: Adventures on the Miskito Shore*, a novel by Ephraim George Squier, a nineteenth-century traveler and authority on Central America. Squier had been President Zachary Taylor's chargé d'affaires, sent by Taylor to Central America to help discredit the English, who were hoping to construct an interoceanic canal along the Rio San Juan. Squier's job was to loosen the affections between the English and the Miskito, bonds forged by earlier generations during the buccaneer era. *Waikna* was the story of an American who traveled with two Miskito Indians in a dugout sailing canoe from Bluefields to Cape Gracias á Dios, along the Miskito Coast.

Nietschmann read the book almost in one sitting; the geopolitical intrigue combined with tropical cultures and natural history was just what he was looking for. By the time he finished the book, he had made up his mind. He would go to Nicaragua and retrace the route Squier had followed from Bluefields to Cape Gracias á Dios, stopping at the Miskito settlements and studying their subsistence economy and culture. He would learn from the Miskito people how they found the offshore shoals and caught the green turtle and how, as turtle fishermen, they were adapting to living with a "fading species."[5]

A Visit to Tortuguero

As soon as he could manage it, Nietschmann headed for Tortuguero and the research station Parsons had described in his book. When Nietschmann got there in August 1967, the turtle tagging was in full swing. It was the height of the nesting season, and Carr and his students were very busy. But he made time for conversations with the young geographer, who understood the ecological and cultural richness of the Central American highlands, an area Carr had grown to love during his teaching stint in Honduras.[6] And Carr was happy to encourage a geographer with an activist's bent to pick up where Jim Parsons had left off. Perhaps Nietschmann could figure out how to end commercial exploitation of the green turtle.[7] At the very least, if the young geographer went to live among the turtle fishermen of Nicaragua he could encourage them to return the tags and claim the reward.

Carr knew exactly why Nietschmann wanted to retrace Squier's route. He himself had made a similar trek through the forested lowlands of eastern Nicaragua while serving as chief hunter for a month-long timber scouting cruise up the coastal rivers. He had published part of his journal in his book *High Jungles and Low*. The trip had embarked from the tiny Nicaraguan port of Bluefields and visited Pearl Lagoon. Both places had been capitals of the ersatz Mosquito Kingdom the English created in the seventeenth century as a bulwark against the Spanish and a haven for buccaneers, and Squier had been to both.

Carr had just updated Parsons' map of the tag returns for his book that was due out next month. For this natural history of sea turtles, Carr had chosen a particularly appropriate title, *So Excellent a Fishe,* a quote from the 1620 turtle law whose text would be reprinted in the frontispiece. Parsons' archival research had confirmed what Carr had always suspected: the history of commercial turtling in the New World began in Bermuda at the end of the sixteenth century.[8] After establishing a permanent settlement there, the English adopted what Carr believed was the very first conservation measure in the New World. The Bermuda Assembly enacted "An Act Agaynst the Killinge of Ouer Young Tortoyes," to stop the local fishermen who "snatch & catch up indifferentlye all kinds of Tortoyses both yonge and old little and great and soe carrye awaye and devoure them to the much decay of so excellent a fishe." But the ordinance had been in vain; Bermuda's green turtle population was one of the first to disappear.[9]

Carr told Nietschmann that the new map would confirm that more than half the green turtles nesting on Tortuguero were from Nicaraguan waters, where the turtles were exploited heavily by the Caymanian turtle fishery, an industry Nietschmann would need to understand for his research on the Miskito turtle fishermen. The Miskito had taught the Caymanians how to find the turtles and now were hired by the schooner captains to watch over the live catches during the turtling season.

Since at least the 1840s, Cayman Islanders had been making two expeditions from Grand Cayman Island to the turtle fishing grounds off the Miskito Coast, first from January to March or April and again from July to September. Excellent mariners, the Caymanians built their turtling schooners themselves on a design adapted from the schooners that fished the Grand Banks for cod off the Canadian Atlantic coast. Their trips lasted from four to seven weeks, sometimes longer.

Upon reaching the Miskito Bank, the schooners would scout out the known turtle resting places, which they could see in the clear Caribbean waters, and mark them with a buoy. They would then send two crewmen out in catboats built like the Grand Banks dories to set large-mesh nets over the sleeping rocks. The turtles would become entangled in the net when they surfaced for breath before beginning their trip back to the seagrass beds. In the morning, the catboat crews would return and haul in the turtles and deliver them to the schooner.[10] They were then unloaded into pens built on the cays until a full load was ready for transport to Key West.

The Miskito lived out on the cays in tiny huts built out over the water to escape the flies. Occasionally the Caymanians on the schooners would buy from the Miskito a live turtle, some dried calipee from a turtle butchered in the village, or the shells of hawksbill turtles.[11] But Carr told Nietschmann the fishery was changing again as the number of turtles dwindled. One by one, the schooners were converting to diesel power so

FIGURE 12 Miskito fishermen landing their green turtle catch at Tasbapauni village, eastern Nicaragua.

Courtesy of Judith Fitzpatrick.

FIGURE 13 Catboat from the *A. M. Adams* ferrying turtles to enclosures, or kraals, at Miskito Cay, thirty miles off northeastern Nicaragua, where turtles were stored before being transported to Key West for sale to Thompson's cannery.
© Wright Langley Archives.

that they could get to the Miskito Cays faster. The sailing times had to be faster to allow for the longer fishing time needed for each captain to fill his hold.[12]

Nietschmann was fascinated to learn that Carr's tag reward program had become a regular channel of communications with both the schooner captains and the Miskito villagers. Carr told him that the letters accompanying the returned tags were so entertaining and informative he was including several of them in the new book, in a chapter entitled "Señor Reward Premio," the name some of his correspondents mistakenly called him. One frequent correspondent was a seminary student in Texas who spent summers at the Episcopal mission stations along the Pearl Lagoon. Harry Neeley had told Carr that in the village of Tasbapauni, between 200 and 300 turtles were caught each year. Many of these were taken

FIGURE 14 Unloading green turtles into kraal at Miskito Cay, off eastern Nicaragua.
© Wright Langley Archives.

near the dozen or so cays located 3 to 15 miles due east of a place called Set Net Point, named for the fishing method the Miskito had learned from the Caymanian fishermen. Carr wanted to know how the turtles found their habitual sleeping rocks near the cays, but if Nietschmann wanted to learn how the Miskito found the turtles, Tasbapauni might be a place he could do his field work.[13]

While explaining the tag returns, Carr told Nietschmann that his most reliable source was Allie Ebanks, captain of the *A. M. Adams*, a Caymanian schooner owned by the turtle cannery in Key West that had always caught the largest number of green turtles. In his letters returning the tags to Carr, Ebanks often expressed concern for the future of green turtles. He thought the people who shipped turtles taken from Tortuguero via the ports in Limón and Colón, Panama, were ruining the fishery. Ebanks especially disapproved of the schooner captain who carried egg-laden female greens across the Caribbean to the cannery in

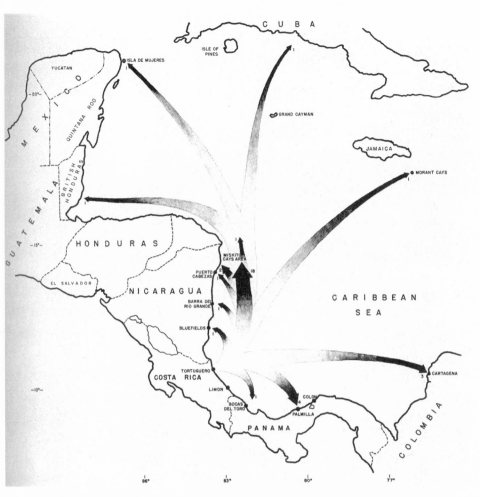

Archie Carr's initial map of flipper tag recoveries from Tortuguero, Costa Rica, 1955–1959.
Courtesy of Carr Family Trust.

Key West. The voyage was one day shorter, but many turtles died in the process. Ebanks was proud of the fact that, while he took twice as many turtles from Nicaraguan waters, he lost none during the passage to Florida.[14]

Carr told Nietschmann that even the diesel-powered schooners were on their way out. There was now a market for calipee, the gelatinous substance that lined the inside of a green turtle's plastron, the belly plate. It was no longer necessary to ship the entire turtle; a poacher at Tortuguero could remove this substance in a matter of minutes and leave a turtle dying in the surf. Also, markets for new products were cropping up everywhere. Just the other day he had heard that a Hollywood actress was now marketing a face cream made with turtle oil; the demand for exotic leather was driving the slaughter of olive ridleys on the Pacific coast of Mexico. Hawksbill shell trade was even on the rise, the availability of plastic substitutes not having sufficed. Carr no longer believed

FIGURE 15 *A. M. Adams* turtle schooner and catboats tied up at cannery wharf, Key West, Florida.
© Wright Langley Archives.

FIGURE 16 Green turtles on the *A. M. Adams* deck at Thompson's cannery, Key West.
© Wright Langley Archives.

that once science had worked out the six big questions on the mechanisms and range of turtle migrations, international legislation would be forthcoming to prevent the slaughter.[15]

The Caymanians' proud tradition of turtling was coming to an end. If Nietschmann wanted to know how they were facing their demise, he should keep a lookout for an article Peter Matthiessen was writing for the *New Yorker*. Matthiessen was both an amateur naturalist and cultural anthropologist. He had written to Carr for advice when he was trying to arrange to sail on the last sail-driven turtle schooner, to write an account of a vanishing profession and a vanishing species. But by the time he was able to arrange a trip with Captain Allie's nephew, Cadian Ebanks, his ship, the *Lydia Wilson,* had also been converted to diesel.

FIGURE 17 (*Top*) Unloading green turtles from the *A. M. Adams* at Thompson's cannery, Key West. (*Bottom*) Hauling green turtles out of turtle kraal at the cannery.
© Wright Langley Archives.

The turtling tradition was dying out, and Grand Cayman was on its way to becoming a tourist haven.[16]

But dwindling turtle numbers were not the only reasons the fishery was on its way out; the Cayman schooners were losing their access to the turtle fishing grounds. At first, it looked as if the fishing rights would continue on the Miskito Bank. Eight years before, Ebanks had told Carr that he had just gone to Managua, the capital of Nicaragua, to renegotiate the treaty that gave the Cayman Islanders license to take turtles in Nicaragua's territorial waters around the Miskito Cays. He wrote: "For your information and interest a delegation from Cayman went to Managua on April 20th. Our mission was to get a renewal of the treaty. We had a very friendly meeting with the ministers of that country. They all agreed that they could not make it 20 years but could make it 10 years. Everything was left for the two governments to draw up in documents.

FIGURE 18 Tourist family watches unloading of green turtles from the *A. M. Adams* at Thompson's cannery, Key West.
© Wright Langley Archives.

The British Ambassador was with us at all times. So I think he is looking after the business end of it."[17]

In the same letter, Ebanks had told Carr that the Cayman Islands government itself was developing a plan "to raise young turtles on one of the islands. We would have to get the eggs and hatch them out and then protect them from birds and fish until they could take care of themselves, which I think is a good idea."[18] Evidently, the government assumed that eventually Nicaragua would not renew the treaty and was looking into alternative ways to make a living from sea turtles.

Nietschmann asked Carr what he thought of the idea of farming green turtles. Was the idea ecologically sound? And what about the economics: would the Cayman Islanders accept it as an alternative to fishing if the treaty was not renewed? Carr told him Ebanks was now retired and would probably not bother taking up farming, and he doubted the younger men would either. As long as their vessels were still seaworthy and there was any market for turtles, they were more likely to shift their fishing to the turtle grass beds off Honduras or somewhere else along the migration route.

Besides, farming was risky because of the frequent hurricanes. And it might take a long time to grow turtles to market size and even longer to grow them to the point where they would lay eggs. Fishermen were risk takers but not particularly patient ones. But the question was definitely intriguing. Anyone familiar with turtle hatcheries at nesting beaches probably thought at one time or another that farming might be the only long-term solution to turtle overexploitation.

Carr then told Nietschmann about Bob Schroeder's idea and about how, when Schroeder visited Tortuguero to take the photographs for the June 1967 article in the *National Geographic*, he, Carr, and Carr's graduate student David Ehrenfeld had debated the very same questions Nietschmann was now asking. He told Nietschmann that when Hurricane Flora struck his pilot farm on Great Inagua Island in the Bahamas the day after the hatchlings were delivered, the turtles escaped from the enclosed seagrass pasture but, remarkably, they had stayed in the vicinity.[19]

Carr explained that despite the turtles' evidently strong site fidelity, Ehrenfeld had been deeply skeptical of turtle farming; raising turtles to be food for people just made no ecological sense to him. And philosophically, he believed endangered turtles should not have to earn their survival through commercialization. But Schroeder was already working on raising a breeding stock and actively seeking investors and a suitable location for a turtle mariculture facility. Carr had mentioned that another site with a natural seagrass meadow like Great Inagua was North Sound on Grand Cayman Island, the same place where the turtle schooners used to pen some of their catches and lay up their boats for maintenance. But, Carr told Nietschmann, he was still of two minds on the subject. Something had to be done to either dampen or saturate demand for turtle products. He had thought enough of the farming idea that he had ended his new book with a discussion of it.[20]

When the Turtle Collapses

Nietschmann worked out that the treaty Allie Ebanks wrote about to Carr was the same one Parsons had described in *The Green Turtle and Man*. An agreement that was an on-again, off-again affair throughout the twentieth century, it first came about after a dispute in 1905 between Nicaragua and Britain. Twenty-three schooners from the Cayman Islands were in the habit of each taking as many as 200 turtles a season, in violation of Nicaragua's three-mile maritime limit claim. After lengthy negotiations, the two nations signed a treaty in 1916 granting the Cayman fishing vessels access to the turtle banks. In return, they were required to stop and register at Cape Gracias á Dios or another port before they began fishing. Then, just before departing, they were supposed to call again at the port and pay the customs officials a modest tax on each turtle they were taking from Nicaragua.

Under this arrangement, between 2,000 and 3,000 turtles were taken each year, according to Nicaraguan official reports. But the new turtle schooners built in the 1930s were fast and efficient. In 1956 alone, the Cayman Islanders exported over 4,000 turtles to Great Britain. In all likelihood, they were not reporting all of their catches to the customs men.[21]

Nietschmann also learned that the Miskito and the Cayman Islanders were closely linked by both blood and occupation. The schooner captains sometimes loaned the Miskito an extra catboat, which they then used to set nets over the sleeping rocks as the Caymanians had been doing for the last century. The Miskito's principal job, however, was to live out on the cays during the turtling season and mind the turtles until it was time to load them onboard the biggest of the schooners for the trip to Key West.[22]

Until the 1950s, the green turtles caught on Miskito Bank were still being shipped live by schooner to Key West. Norberg Thompson, the owner of the Key West cannery, hired boat builders on the Cayman Islands to build several large schooners. But by the time Nietschmann learned of the trade, only the *A. M. Adams*, the one Thompson had named for the general manager of his cannery, was still working. Advances in freezing technology had made it possible to freeze meat from the large numbers of green turtles slaughtered in factories in ports like Limón and Georgetown. Soup companies in Europe were now able to avoid the shipping costs by buying frozen or dried turtle meat and calipee. Nietschmann had seen evidence of this change with his own eyes when he visited Tortuguero. Early one morning, he had stumbled across a green turtle a poacher had sliced open the night before to remove its calipee and had left to die in the surf zone.[23]

Carr was right; things were definitely changing in the green turtle economy in Nicaragua, and Nietschmann had better get there soon if he wanted to see things firsthand. Nietschmann would go to Tasbapauni and learn how the villagers hunted green turtles and how their turtle-based subsistence economy was changing. But he was still thinking about what he had pondered that night on the beach at Tortuguero: who would kill the last turtle? Would it be one of the fishermen from Limón hovering three miles offshore exploiting the loophole in Costa Rica's turtle law? Would it be a shark swimming off the beach waiting to make a meal of the last nest of hatchlings, a Miskito Indian for food, or a turtle company for meat and leather?[24]

As soon as he could arrange it, Nietschmann, his wife, and their young son moved to the Miskito Coast. Judi Nietschmann was an anthropology student studying health-related rituals. When the family got to Nicaragua in September 1968, Nietschmann found out that the turtle schooner voyages had ended abruptly. One day the customs officials just told the captains, when they came ashore to pay their duties, that Nicaragua had not extended the treaty as Captain Ebanks had expected. The government had tired of the schooners' habit of evading the customs they were supposed to pay at Cape Gracias á Dios.[25]

FIGURE 19 Green turtle dying in the surf at Tortuguero, Costa Rica, after a
turtle hunter has cut out the calipee, the gelatinous substance used to make
soup, from above the turtle's plastron.
Special Collections, George A. Smathers Libraries, University of Florida, Gainesville,
FL, courtesy of Carr Family Trust.

The last several decades of turtle turning at Tortuguero had reduced
the numbers of turtles on Miskito Bank, but there were still enough to
support the Tasbapauni subsistence economy of the village, which re-
volved around the green turtle. The Nietschmanns took up residence
in the village, and Barney set about interviewing the fishermen on how
they found the turtles. He built his own sailing canoe with help from a
villager, Baldwin Garth, and then sailed with him to retrace the route
Squier had taken to Cape Gracias á Dios. After the trip, Nietschmann
was invited to accompany the village men on their turtle hunting trips.

The navigational and fishing prowess of the men of Tasbapauni was
impressive. Nietschmann learned that the men of the village utilized an
area of marine waters of approximately 600 square miles. Within this area
were twenty major turtle banks and almost forty important shoals. They

FIGURE 20 Nicaraguan customs officials collecting turtle fees from Cayman Islands turtle schooner captain at Cape Gracias a Dios, Nicaragua. © Wright Langley Archives.

would follow the green turtles' predictable patterns and capture them with harpoons while the turtles fed on the grass banks or made their shoal-to-feeding-grounds commute. To find these shoals, each turtle-man carried with him a mental map of his village's area and the named shoals and banks within that area. The men would paddle their twenty-foot sailing canoes through the breakers in the dark and then catch the evening land breeze to sail. They used the stars to guide them eastward and found the turtles' sleeping shoals and the feeding banks through the characteristic surface swell motion of these areas.

Nietschmann believed that the turtle fishery was sustainable because only a few men in the village had this special sea-finding ability. This limitation and the difficulty of harpooning meant that turtles were taken one at a time by two men, and there was little risk of overexploitation. The turtle meat was shared and sold in the village, and this practice served to bind the community as well as meet its nutritional requirements.[26]

But the sustainable level of exploitation was changing rapidly. The village men were now using nets to catch the turtles, and this process required much less skill. The Miskito fishermen had gotten nets from the Cayman schooner captains, but net-fishing had never been a large part of their livelihood. Now, the nets were distributed extensively to the villagers on credit by a foreign-owned turtle company that opened up in Puerto Cabezas in late 1968. A second plant, Frescamar, was opened in Bluefields in 1969. Like the Cayman schooner turtlers, the village fishermen would set the nets in the middle of the day, when they could look down and see the dark shoal areas; the nets would be hung vertically in the water, anchored to the bottom by one line and buoyed with floats. Netting turtles required much less skill, allowing just about anyone to enter the fishery.

The Miskito were now setting thousands of these fifty- by fourteen-foot nets daily over the turtles' nocturnal resting habitats. The nets drifted in the currents and snagged the turtles as they returned to the shoals for the night. No apprenticeship was necessary to learn the skill—all they

FIGURE 21 Green turtles from one day's 300-turtle shipment received from Miskito Cay awaiting slaughter at the Frescamar turtle factory, Bluefields, Nicaragua, 1972.
Courtesy of Carr Family Trust and University of Florida, Archie Carr Center for the Study of Sea Turtles.

Who Will Kill the Last Turtle? 127

needed to do was to follow the older men who knew how to find the shoals and then set the nets and retire for the night. At dawn, they could remove the entangled turtles, take the turtles to the pens, and wait for the company boat to pick them up.[27]

Each turtle slaughterhouse had the capacity to process 10,000 turtles a year. To encourage the Miskito to catch that many turtles, the companies gave the villagers building materials to make houses on the offshore cays. This allowed them to stay out on the turtle banks during rough weather and get in more fishing time. The companies then sent boats up and down the coast on weekly runs, bringing the Miskito fishermen food and more fishing gear, and buying their turtles for cash.

The company boat would visit frequently to keep the fishermen from returning to their villages with the turtles, as they had done in the days of the Cayman schooners, when the schooners often did not return for

FIGURE 22 Villagers vying for a chance to buy turtle meat at Tasbapauni village, eastern Nicaragua, from increasingly rare green turtle landings. Turtle-freezing factories were purchasing the majority of the villagers' catch. Courtesy of Judith Fitzpatrick.

weeks at a time. When the men did return home, they brought few if any turtles. The vast majority were sold to the company for cash. The village fishermen no longer liked bringing a turtle home for distribution: they sold 90 percent of their catch to the companies, so there was not enough now to fulfill kin and social obligations, nor to meet the village's nutritional needs. There were endless problems deciding to whom to distribute the turtle meat, whether by gift or by sale. Some fishermen avoided the problem altogether by selling all their catch to the companies. Social tension was growing in the village, and the villagers' diet suffered. The outside sources of their food were subject to world market fluctuations and inflation. A sense of being poor was beginning to overtake the community.[28]

As soon as Nietschmann got back to Wisconsin, he wrote to Archie Carr to tell him about the turtle factories and their impact upon the people of Tasbapauni. As he wrote his dissertation about the Miskito villagers and their turtles, he remembered the evening he sat on the beach at Tortuguero and wondered who would kill the last green turtle. He now knew the answer. It would be the workers in the turtle slaughterhouse in Bluefields or Puerto Cabezas. What a paradox that they would kill off the last turtle only a few dozen miles away from Tortuguero, where that turtle had been made possible by the remarkable conservation program designed to ensure the species' survival.

CHAPTER NINE

Red Data for the Green Turtle

The 1967 field season at Tortuguero was Archie Carr's longest to date. When he returned to Gainesville, Carr faced a mountain of correspondence and neglected paperwork. He felt a tinge of regret that he had spent so much time to see so few turtles return to nest. His most overdue project was to prepare the list of sea turtle species that were threatened with extinction. When Peter Scott had invited Carr to chair the Survival Service Committee's Marine Turtle Specialist Group in 1964, he had made it clear that this was the group's principal function. Carr had promised to canvas the members for information on the status of populations around the world, put the information in the Red Data Book's format, and send it to Scott for inclusion in the book's first volume on reptiles and amphibians.

But Carr had difficulty applying IUCN's endangerment classification scheme; he found it especially hard to compare each species' predicament. For the past year or so he had been writing a book of his own on sea turtles and wanted the last chapter to discuss the future prospects for sea turtles: which species were likely to go extinct and which might hang on if protective actions were taken immediately. But writing the chapter had been a challenge. Sea turtle taxonomy was a puzzle, given the number of distinct populations and nesting colonies, any one of which might be an incipient species. To develop a strategy that could prevent a particular species' extinction required more scientific certainty than he currently had. Under these circumstances, Carr thought all nesting grounds of all sea turtles should be protected; the current rate of growth in human populations in coastal areas meant turtle egg harvesting was simply unsustainable. And, he thought, "if it is the long run you think of, *all* sea turtles are endangered."[1]

Furthermore, commercial exploitation of turtles was at an all-time high. Carr's son had told him about seeing the remains of hundreds of

Pacific ridley (*Lepidochelys olivacea*) turtles piled high behind new slaughterhouses constructed next to the recently discovered mass nesting beaches on Mexico's Pacific coast.[2] Carr thought that if the Pacific ridley was so abundant that one million or more could be slaughtered for only a few strips of leather, it might be hard to argue that the species was on the verge of extinction. If this trend continued—and there was no reason to think it would do anything but continue—extinction was only a matter of time. But if the Red Data Book classification was a matter of pure numbers, most species were not approaching extinction.

Carr thought that only one species seemed headed for imminent extinction in terms of numbers. The Atlantic ridley (*Lepidochelys kempii*) had only one nesting population—on the beach at Rancho Nuevo, Mexico. The site of its *arribada*—mass nesting event—had been pinpointed just a few years before when Henry Hildebrand, a biologist at the University of Corpus Christi, found a home movie shot in the 1940s by a young Mexican engineer and pilot, Andrés Herrera. Hildebrand's discovery had solved the taxonomic "riddle of the ridley" that Carr had written about in *The Windward Road*. The Atlantic ridley was indeed a distinct species and not a hybrid, but it was now nesting in numbers closer to a dozen, not the 40,000 that Hildebrand estimated from Herrera's film.

Carr thought that if one looked at the trends and the rates of exploitation, many species in addition to the Atlantic ridley seemed to be very much at risk. He had once thought that "tortoiseshell" combs and eyeglass frames made from plastic would eliminate the market for hawksbill shells. But plastic did not satisfy the affluent fashion-seekers of the 1960s, and the demand for these luxuries was going back up. The calipee, leather, and shell a poacher could take from an adult hawksbill put a price on its head that was too hard for fishermen to resist. The tourists who were beginning to flock to the Caribbean were happily buying stuffed and polished yearling hawksbills as souvenirs.[3]

To buy more time to complete the Red Data Book entries, Carr wrote to Peter Scott:

> At the moment . . . the main problem in sea turtle survival is not just deciding which "species" are in peril, but rather fighting to provide every scrap of protection possible for all populations of all the five genera. The egg markets of the Sarawak Islands and those of Eastern

Malaysia ought to be stopped. So should the calipee trade; the expanding commerce in turtle skins; the Japanese exports of stuffed green turtles to be used by morbid Californians as household furnishings; and the worldwide traffic in young hawksbills, polished and mounted for hanging on the wall. The aimless harpooning and netting of sea turtles by commercial and sport fishermen should everywhere be illegal. And so on. . . . So it seems to me that all sea turtles are in grave danger, and that the Mexican ridley is just the first extinction to expect.[4]

Dangerous Categories

When Carr was finally able to sit down and write the Red Data Book pages on *Chelonia mydas*, the green turtle, he endeavored to fit the information he had into the Survival Service Committee's headings: present and former distribution, status, estimated numbers, breeding rate in the wild, reasons for decline, protective measures taken, protective measures proposed, number in captivity, and breeding potential in captivity. To do so, he drew mostly on his own work in the Caribbean, but he also quoted from the reports of Tom Harrisson for the South China Sea and Harold Hirth for the populations in the Indian Ocean.[5]

There was no specific category for "rate of exploitation," but under "estimated numbers" Carr quoted from Jim Parsons' exhaustive global survey of turtle use in his book *The Green Turtle and Man*. Even though it was now almost ten years old, Carr believed this information made clear that in every part of the tropical world, traditional green turtle fisheries were being transformed into a production line for the turtle soup manufacturers in England and North America. For "estimated numbers" Carr wrote:

> Unknown and almost impossible to ascertain. The Green Turtle is a vegetarian and browses on marine grasses in tropical shoal waters. Almost their entire lives are spent at sea, the females alone leaving the water and then only at intervals of several years. Harrisson and Hendrickson consider that the 1–2 million eggs taken each year from the Sarawak turtle islands may represent the production of from 2–4000 females and that there may be 10,000 females in that

particular community. Another million eggs are collected annually from the Philippine turtle islands. Although there are no statistics, the number of eggs recovered from Malayan beaches is said to run into millions. Large quantities find their way to Singapore markets where, in 1961, eggs were selling at £1 per 100. Between 15 to 20,000 live turtles are marketed commercially in North America and Europe each year. Until recently the Bajun Islanders of East Africa were content to catch sufficient for their own requirements but, since 1950, the export of live turtles (taken with the aid of sucker fish, *Remora remora*) has become an important industry. 1000–1500 live turtles (perhaps half the Bajuni catch) are exported annually. Green turtles are now Royal Game in Kenya and their capture is controlled. 2500 per year are killed along Turkish coasts.

For "reasons for decline" he wrote: "The very localized feeding and nesting grounds and the turtles' size have made it extremely vulnerable to man. . . . In most parts of the world it is prized for its flesh but in South East Asia the meat is seldom eaten although the eggs are systematically harvested. Expanding human populations have resulted in excessive exploitation of the turtles and its elimination from many breeding grounds."

Under "protective measures taken" Carr described Harrisson's pioneering work on the turtle islands off Borneo in transplanting many thousands of eggs into special hatcheries, "(100,000 in 1962, for example) for subsequent release into the sea."[6] He also mentioned the nest transplantation projects in the Pacific Trust Territories, adding that the Caribbean Conservation Corporation, under his direction, had taken the idea one step further by making available free stock from the Tortuguero hatchery "to any agency able to ensure protection to nesting beaches formerly used by Green Turtles, but now abandoned by them. In the summer of 1960, and each summer since, some 15,000 have been hatched and flown to different parts of the Caribbean for release. The . . . hatch averages more than 70%. . . . The only advantage of hatchery rearing is circumvention of predation on eggs and on hatchlings crossing the beach and entering the surf."

Carr gave most of his attention to the item the Red Data Book called "protective measures proposed." Here was his opportunity to lay it all out: there could be no half-measures if IUCN or any other organization truly wanted to protect the remaining populations of green turtles. In his view, six actions were needed to prevent the green turtle's extinction. The first three involved getting more information—on the turtles and on why they were being hunted. These actions were (1) range reconnaissance, to fill in the serious gaps in known nesting and feeding grounds; (2) surveys of present legal and illegal exploitation of the green turtle and the kinds of products turtles were being killed for; and (3) studying migration and population by introducing tagging at all the major nesting grounds.

Carr's fourth action addressed what he knew was the cause of the green turtle's decline—international trade to supply the soup and leather industries. He knew that no one nation could stop this trade on its own, so his fourth recommendation was the initiation of "negotiations for the international control and regulation of all trade in sea turtle products, which should be reinforced by propaganda campaigns to depopularize their use." Now, he thought, was when Tom Harrisson's ideas should be put into action—to make people think twice, maybe three times, before they ever ordered another bowl of turtle soup. While he agreed whole-heartedly with Harrisson that this had to be done, he knew that other members of the Survival Service Committee might take some convincing. Carr added a bit of explanation: "The use of turtles as human food . . . and as an instrument for raising the food production of the oceans, is a legitimate and worthy aim that simply cannot be realized because over-use has grievously damaged the resource. For the time being, thus, there seems no recourse except strict curtailment of all commerce in products derived from marine turtles."[7]

Carr knew and admired many people in the tropics who depended on green turtle eggs and meat as a source of protein. These included the Miskito Indians of Nicaragua, whose turtle subsistence fishery he had encouraged Barney Nietschmann to study, and Mrs. Ybarra, the Tortuguero woman he had met on the black beach and had made famous in the *Mademoiselle* short story that had won him the O. Henry prize in 1955.[8] Perhaps because a ban on the use of wild green turtles would hurt these people, and because he knew how delectable green turtle soup was, Carr added a fifth recommendation for the Red Data Book pages: turtle farming. He wrote:

(5) Pilot Culture Projects. Experimentation in the rearing of Green Turtles in enclosures should be instigated and supported. If there is any promise in the sea turtle industry it will only be realized through turtle farms. As a first step toward the development of these, a pan-tropical search should be made for localities with (a) good sites for hatchling crawls, with warm, clear constantly renewed water of high salinity, (b) extensive areas of submarine spermatophyte vegetation for pasturage or cut feed for the maturing stock, (c) a constant supply of cheap animal food for the first-year turtles, and (d) freedom from typhoons, hurricanes and drastic changes in water level. To insure the supply of hatchlings for such projects the problem of inducing the migratory Green Turtle to mate and nest in captivity should be studied. There seems no inherent reason why *Chelonia* should not become a semi-domesticated meat animal of great value, supported by aquatic pasturage as cattle are supported by terrestrial grass. Successful evolution of such culture would not only extend the means of taking food from the sea but would quickly take the pressure off of wild sea turtle populations, and thus help save the species for the distant future.

Carr based his sixth action on his recent correspondence with Harold Hirth. Hirth had put so much work into Operation Green Turtle and the hatchery at Tortuguero that Carr chose his photograph of tagged hatchlings for the cover of *So Excellent a Fishe*. Carr recommended that the Aldabra atoll in the Seychelles be set aside as a permanent nature reserve and that green turtles be given total protection throughout the archipelago. This protection should endure for at least ten years and include "some sort of international regulation" to prevent "foreign" ships from taking turtles there.

Carr's hope in writing the turtle farming recommendation was that a global organization like the UN Food and Agriculture Organization or IUCN would provide funds for pilot turtle farming projects in places where subsistence turtle fisheries and egg collecting needed to be banned. Domestication was really the only plan for green turtle exploitation that he could support. Ten years before, Carr had agreed with John Hendrickson's conclusion that managed egg harvests combined with bans on hunting adult turtles could be sustained.[9] But that was before the calipee trade bloomed and the new turtle products became fashionable.

Carr now believed that exploitation of wild turtles could never be effectively managed for a sustainable yield. As long as people insisted on having turtle products, only domestication could allow the green turtle to be saved. He knew mariculture had problems, and he vividly recalled David Ehrenfeld's arguments to Bob Schroeder that it made no sense ecologically to raise green turtles for food by feeding juvenile green turtles other marine species. People could just as easily eat those species without the energy costs and other expenses of converting fish to turtle meat. But Carr thought that if the objective was to save wild turtles from the slaughterhouse, raising young green turtles on fish meal and other fisheries by-products might make some sense. Green turtles are important components of tropical marine ecosystems. And cultivating turtles was the only way to convert the sunlight and protein in the seagrass beds into food for humans.[10]

Carr did not want to condemn all utilization of green turtles; it was the excessiveness of their exploitation and the purpose—for luxury and status items rather than necessities—that he could not abide. He had made his view very clear in the final chapter of So Excellent a Fishe, from which Carr distilled his Red Data Book pages. The natural history of sea turtles made them too easy to catch, and the population boom in the coastal tropics made it likely that consumption of turtle eggs and adult turtles would continue to increase until they were all gone. They were just too good to last. This was the message he took from the disappearance of the Atlantic ridley arribada at Rancho Nuevo and, he feared, from what he was seeing in the nesting decline at Tortuguero.[11]

A Meeting at Morges

Carr was later dismayed to learn that the Red Data Book is not the place to lay out an international recovery plan. In May 1968, Peter Scott and Colin Holloway, the executive secretary of the newly renamed Survival Service Commission, convened a meeting at IUCN headquarters in Morges, Switzerland, to review the draft Red Data Book pages submitted by the new species specialist groups. Carr did not attend; he had chosen instead to go to IUCN's meeting in Argentina the month before, where he had spoken with Holloway about the proposed actions, and he could not afford the additional time to go to Morges. He knew Tom Harrisson would be there to make sure the turtle action items were not watered down. Scott had recently suggested that Carr appoint Harrisson as co-chairman of the Marine Turtle Group. Carr had readily agreed. He was already overcommitted and did not especially like contentious discussions and endless meetings. Harrisson seemed not to mind all the meetings and was frequently at IUCN headquarters in Morges for meetings of the Survival Service Committee's Alert Group and in connection with his work with Barbara Harrisson on orangutans and other disappearing species in Southeast Asia.[12]

Despite Harrisson's presence at the May meeting, Carr's policy recommendations were pared down from six to three. All that remained of his "protective measures proposed" were the nature reserve on the Aldabra atoll in the Seychelles and the pilot culture projects.[13] But Harrisson had managed to convey the seriousness of the global situation for marine turtles and to convince Colin Holloway to find funds to convene a meeting of sea turtle specialists that could begin to tackle the problem.[14] Herbert Mills, from the World Wildlife Fund's U.S. National Appeal, had offered to fund the meeting so after much correspondence back and forth on details, the meeting was finally arranged to be held in Morges in March 1969.

Carr hoped this meeting would be an opportunity to secure adoption of serious protective measures for endangered sea turtles. He knew this would take persistence and a presentation of all the facts as they were known, and he agreed to chair the meeting only after Sidney Holt from the FAO turned down Holloway's invitation. Carr had wanted someone else to run the meeting so he would be free to participate fully and to convince Holt, as head of the FAO's fisheries division, to get directly involved with the overexploitation of the green turtle. The Tortuguero green turtle population was in dire straits; despite the best efforts of the Brotherhood of the Green Turtle and Carr's partners in Operation Green Turtle, population enhancement projects had proven to be insufficient to stem the losses from harvesting. It was time for prohibitive policies.

Before the meeting, Carr and Harrisson suggested a list of people Holloway should invite. Some were members of IUCN's Marine Turtle Specialist Group, had sent Carr data and information on their turtle populations for the Red Data pages, and were clearly appropriate. Others were not so clear. Leo Brongersma, a Dutch zoologist and director of the Netherlands Natural History Museum, had limited field experience with sea turtles, given the paucity of marine turtles in Europe. But he wanted to attend and had strong views on how species should be classified by conservation status. Harrisson was not so sure, but Carr agreed with Holloway that as a member of the specialist group, Brongersma should attend. Because he did not represent a country that harbored any sea turtle populations, he would be listed on the agenda as an observer, and he would not get any of the limited travel funds reserved for national delegates.[15]

Most of the sea turtle experts were on the invitation list because Holloway and Harrisson hoped they could represent or influence their country's policies in favor of conservation. Carr agreed that at this stage of the game, having such influence was even more important than zoological expertise. For example, a Scottish zoologist named Robert Bustard, who studied Australian lizards, had begun to study green turtles nesting on Heron Island on the southern Great Barrier Reef. This was the same nesting beach where Frank Moorhouse had done one season of tagging in the shadow of a turtle cannery. When Moorhouse realized that the female green turtles were repeat nesters but were being captured before they had laid their first set of eggs, his recommendations led to the first sea turtle conservation legislation ever enacted in Queensland. In 1932, the government banned turtling during the early part of the nesting season, to give the females a chance to lay at least one clutch before being killed. Carr thought Bustard's research, in a sense, reinvented the wheel that others had built, but he agreed that Bustard should be invited.[16] If legislators in Queensland were that responsive to a scientist's recommendation, there was hope that more could be done in that state, whose waters harbored such important turtle populations.[17] When Harrisson wrote to Bustard to invite him, Bustard accepted with enthusiasm and offered to represent all the areas of Australia known to have turtles, not just Queensland, including Western Australia, the Northern Territory, and the Australian territory of New Guinea.[18]

Carr still hoped IUCN would spearhead a massive effort on behalf of marine turtles and could convince the FAO to join the effort. After all, turtles, especially the green turtle, were a food resource in the coastal tropics. So Carr's goal for the meeting was to get IUCN to hire a full-time staff member who was able, and not disinclined as Carr was, to travel to meet with governments to press them to enact the necessary legislation and engage in international negotiations to curb the turtle trade.[19]

Carr knew just the man for the job. Oxford-educated Peter C. H. Pritchard was tall, handsome, and utterly charming; had just finished his PhD under Carr's guidance; and needed paid employment where he could use his immense talent for turtle conservation. Pritchard had recently discovered a leatherback nesting colony in the southern Caribbean and had trained the local Amerindian people to guard the turtles rather than hunt them. He convinced chiefs and lawmakers in French Guiana and British Guyana to protect the leatherback nesting beaches under their jurisdiction. At the meeting, the specialists readily agreed young Pritchard should serve as rapporteur. They were impressed when Holloway read a letter and telegram the Survival Service Commission had received from the Guianas announcing that they had adopted the laws Pritchard had recommended. He looked like a shoo-in for the job.

Carr wanted the neighboring countries of Nicaragua and Costa Rica to reach an understanding regarding their shared green turtle population. He knew from his own contacts that Costa Rican officials were reluctant to fully protect the Tortuguero nesting population as long as fishermen in Nicaragua were still free to capture them on the feeding grounds. Carr proposed that IUCN organize and fund a meeting where such an agreement could be reached. For the meeting to be successful, it was important that it not be convened by a group of North American conservationists such as the Caribbean Conservation Corporation, although its staff would be happy to attend as observers. He was very anxious to avoid the appearance that outsiders were meddling with national resource policies.[20]

During the delegates' reports on national and regional situations, Carr was blunt and to-the-point. The sharp drop in the number of nesting green turtles at Tortuguero was a signal of an actual population decline, not merely season-to-season fluctuations reflecting the animals' variations in breeding cycles. He told the group that two new turtle freezing plants had recently begun buying green turtles from the Miskito Indians and were processing turtles by the hundreds every day in Puerto Cabezas and Bluefields, Nicaragua. He asked the FAO observer whether there was any truth to the rumor that the funding for these plants came from the US Agency for International Development or the FAO. Dr. D. W. Sahrhage responded that he knew of no such connection but would contact FAO headquarters in Rome for more definitive information. He welcomed the suggestion that he submit a report of the meeting to the FAO.

In almost every regional report, the delegates described similar problems: the country in which a green turtle nesting beach was located was reluctant to crack down on local hunting or reduce egg harvests as long as that beach was supplying a fishery somewhere, especially if that fishery was in foreign waters. The reports convinced Carr that it was time for the international negotiations he had recommended in his draft Red Data Book pages. And there was no time to lose; exploitation was likely to increase as even more turtle products became available and increased in popularity like the turtle oil cosmetics being marketed in California. He told the delegates it was time for international controls on all trade in turtle products, "reinforced by propaganda campaigns to depopularize their use."[21] But the delegates did not agree. They favored publicizing the plight of sea turtles around the world but were opposed to a depopularization campaign until the survey of exploitation they had endorsed was completed. Several expressed the concern that it might be very difficult to amass the data on exploitation if the companies using sea turtles identified conservationists as enemies.

The newest turtle products—leather and cosmetic oil—might become even more ubiquitous, Carr told the meeting. One million dollars had been invested in a commercial turtle farm called Mariculture, Ltd., and that farm had just been incorporated in the Cayman Islands. Robert Schroeder, the managing director and scientific coordinator of the operation, had gone with his principal investor, an American businessman named Irvin Naylor, to Managua and San José in search of eggs and turtles with which to build the farm's brood stock. The Costa Rican ministry of agriculture had given them permission to take 10,000 green turtle eggs in return for a promise to return and release about 10 percent of the hatch as yearlings on the beach where the eggs were taken: Tortuguero.

Although he did not say so in so many words, Carr was very unhappy about this development. He said that the farm was "cannibalizing" Tortuguero for eggs and that it would soon do the same at the Ascension

Island green turtle colony, where he also had a tagged population.[22] Marlin Simon was the man in charge of collecting the stock for the farm. He knew quite a bit about hatching and raising green turtles; he had worked for Carr on Operation Green Turtle.

Although the first day's topic was supposed to be the life histories of marine turtles, the delegates began an unscheduled discussion of the farming question. Carr told them that Schroeder's farm was not the kind of "pilot culture project" he had in mind when he recommended culture in the Red Data Book's page for *Chelonia mydas*. The other specialists agreed that they would oppose any farm that depended indefinitely on wild stocks for eggs or hatchlings. But if a farm could produce its own stock, they would fully approve.

But, Carr asked, how reasonable was it to expect that captive green turtles would lay eggs in an artificial setting? Had it ever happened? John Hendrickson mentioned that several clutches of green turtle eggs had been laid and hatched on an artificial beach built on the edge of a small, saltwater enclosure at his aquaculture research facility in Hawaii. He, for one, was optimistic about the prospects for entirely captive breeding colonies. Someone else pointed out that in some places, farms could be stocked with no harm to wild colonies by taking those eggs laid early in the season that were usually destroyed by later-nesting turtles.[23]

Later on in the meeting, Hal Hirth gave his report on the situation on the Aldabra atoll. The meeting delegates passed a resolution endorsing the idea that the island be declared a complete nature reserve, where all animals would be safe from exploitation, including sea turtles. Hirth reported that Curieuse Island in the Seychelles had been used in the past as a turtle holding pen and, with renovation, might be a good place for a turtle farm. Carr recommended that any Seychellois that wanted to learn farming methods should go to Grand Cayman Island, since there was "already a one million dollar turtle farm operation" under way there.

When the topic of the meeting turned to international requirements, Hendrickson outlined seven areas of research that he felt were essential: taxonomic identity of all living sea turtles, turtle culture, immunological tagging, where turtles spent the "lost year," leatherback research, turtle flesh poisoning, and the creation of a central tagging agency.[24] The FAO representative, Dr. Sahrhage, said his agency could assist with turtle tagging projects and would be happy to publish, as Carr suggested, public information sheets in several languages explaining the purpose of tagging and the researchers' need for finders to return them. The FAO also planned to publish synopses on all sea turtle species as part of a series on economically important aquatic organisms. Hal Hirth had agreed to write the first one on green turtles. Carr added that population structure should be researched, as should the energetics and productivity of turtle grass ecosystems. Reflecting his recent experience with the insurrection in Borneo, Harrisson suggested that research should focus in places that were politically stable. Carr, Hendrickson, and Bustard agreed to form a committee to set priorities among these research topics. An observer from the Caribbean Conservation Corporation agreed to make a survey of turtle exploitation, uses, and markets around the world.

Tom Harrisson read a statement he had drafted on the world marine turtle situation and a seven-point program for their future conservation. His seven points received approval by acclamation. They were increased hatchery programs, analysis of world exploitation patterns, a broad public information program, identification of all nesting beaches by survey, creation of sanctuaries under scientific management, appointment of a full-time officer at IUCN to work with the turtle group, and further meetings of turtle specialists like the successful one that was concluding.

Carr asked that the following resolution be adopted:

> The members of the conference agreed that the use of sea turtle products is desirable. It was agreed, however, that until clear surpluses are demonstrable local use should take precedence over export, and the specialists were unanimous in their opposition to overexploitation everywhere. Overexploitation was defined as imposition of a drain trending towards exhaustion. To determine when terminal drain is in progress an inventory of its kinds and volume has got to be made. To reverse terminal drain requires legislation and enforcement appropriate to the situation. If legislation fails to stem the drain then a world campaign to depopularize the turtle product most evidently involved in the depletion is regarded as the only course.[25]

There it was. He had finally spelled it out. To his surprise and satisfaction, the resolution was approved without amendment. But he had had to compromise on his desire to see an immediate campaign to dampen the new market for turtle oil and leather.

When the turtle farming issue came up formally on the agenda, Carr read from an article he had published in the January–March 1969 issue of the *IUCN Bulletin*, a synopsis of the concluding argument he wrote in *So Excellent a Fishe*:

> One recourse that appears to offer promise is turtle mariculture. There is no inherent reason why green turtles, for example, cannot be profitably reared through their carnivorous first year and then put out on turtle grass pastures. Two harmful side effects to be expected from the premature spread of turtle culture are the inevitable parasitization of natural nesting grounds for eggs with which to stock rearing pens, and the stimulation of new markets and higher prices before the volume-production necessary to satisfy the demand and relieve the pressure on wild populations is achieved. Controls should be set up to keep turtle farms from prematurely attempting commercial operation. Permits to take eggs should be issued only to sound pilot projects that embody plans for careful efforts to abridge the complex migratory life cycle of the species and induce it to breed in enclosures. Production for profit should go ahead only when results of these experiments make the project independent of natural nesting grounds.

When he was finished reading, John Hendrickson spoke first. He agreed with Carr's statement; the principles for successful farming had to be developed by a noncommercial entity. He was sanguine about the prospects for developing a breeding stock. And if anyone was going to fund an experimental green turtle farm, he knew a good location. He recommended a site at the head of the Gulf of California, not far from his hometown of Tucson, where he was headed for a new job at the University of Arizona.[26]

Robert Bustard was also listening intently. He too had a good place in mind for pilot turtle farms. Some nesting beaches on the Great Barrier Reef were so heavily used by the turtles that many early egg clutches were probably destroyed by later-arriving females. He had simulated this phenomenon in his recent paper on density-dependent population regulation.[27] If there were such beaches in the Torres Strait Islands, off the northern coast of Queensland, where the indigenous people were still hunting turtles, the otherwise doomed eggs could supply cottage-scale turtle farms that could replace the wild turtles that were hunted. And he could train the turtle farmers in proper husbandry techniques that Hendrickson and others had developed and that Robert Schroeder was working out at Mariculture, Ltd. He made a note to look into government funding for experimental turtle farms on the islands of the Torres Strait. Before the end of the meeting, he also made sure to ask Carr for Robert Schroeder's address. He wanted to write him for more about how Mariculture got started and how it planned to use its million-dollar investment.

At the meeting's end, Colin Holloway invited all the attendees to become members of the Survival Service Commission's Marine Turtle Specialist Group and to think about another meeting in two years' time. For the last item of business, he asked the group whether, if funds could be raised, the Survival Service Commission should appoint a full-time executive officer for the marine turtle group. He told the group that Archie Carr had suggested Peter Pritchard for the position. As a point of order, with his European formality Leo Brongersma questioned whether it was appropriate to discuss a person's candidacy while he was in the meeting serving as rapporteur. John Hendrickson said he would like more time to consider Holloway's proposals and to see a proposed scope of work for the marine turtle group and its executive officer. Robert Bustard asked whether the officer would be stationed at IUCN headquarters and live in Switzerland. If so, should not the officer be someone with more experience with national fisheries departments and ministers and a record of successful lobbying for protective turtle legislation?[28]

Although a minority of the delegates seemed to resent that Carr had hand-picked his own student for the new post, Pritchard himself was only mildly embarrassed by the awkward situation and felt relieved when the motion to approve the proposed position was unopposed. Pritchard began planning how he would carry out the seven-point program Harrisson had drafted and Carr's successful resolution on depopularization. He would start by visiting the turtle farm in Grand Cayman as soon as it could be arranged.

A Shaky Start

After the meeting and the opposition to Pritchard expressed by Brongersma and Bustard, Carr began to wonder if he should have invited them to join the marine turtle group and if Holloway should submit their names for appointment to the Survival Service Commission when it met later that spring. A year earlier, Brongersma had asked Colin Holloway if he could become a member of the marine turtle group. When Holloway contacted Carr about it, Carr had readily agreed. But Brongersma seemed convinced, despite the evidence, that marine turtle species could be exploited without endangering their survival.

Carr recalled that Brongersma had suggested that IUCN support moving Africa's sea turtles from their current class A status to class B under the African Convention on the Conservation of Nature and Natural Resources. This would allow African governments to manage egg harvesting on nesting beaches where the turtle was a food staple and the stocks could tolerate the loss. Carr believed that all egg taking and killing of nesting females should be stopped at once. He had made a point of describing in his Red Data Book page for *Chelonia mydas* the overexploitation the Kenyan government had allowed for the overseas soup makers. Now that the green turtle was Royal Game in Kenya this might mean a reduction in this drain, but he still felt that African sea turtles had poor survival prospects. Holloway had forwarded Carr's views on the matter by letter to Brongersma; Carr had stated that he was sure that "Africans in the long run will be better off if they sacrifice a little in an effort to build back their lost and waning colonies than they would be by draining them dry by uncontrolled exploitation."[29]

Brongersma had also suggested, unhelpfully to Carr's mind, that the Survival Service Commission focus on research rather than on matters of international legislation. Yet he was clearly passionate about the fate of marine turtles—in fact, of all reptiles. At the meeting, someone had told a story about how the two Dutchmen present, Leo Brongersma and Joop Schulz of the Suriname Forest Service, had met previously under less happy circumstances, when they were prisoners in the same Japanese internment camp in Indonesia during the war. Brongersma convinced the guards to let the younger Schulz go outside the fence so he could collect frogs for Brongersma's collection.[30]

Carr thought Bustard seemed to be hotheaded, undiplomatic, and too impressed with his own efforts on behalf of marine turtles. With his display of "choleric distress," did he really think he was a better man for the Survival Service Commission job than Pritchard?[31] Bustard had hinted broadly that he knew how to get government officials to act when he told the meeting that it might not be necessary for IUCN to write to the Queensland or Australian Commonwealth governments on behalf of marine turtles. He said that when the chief inspector of fisheries for Queensland visited his research station on Heron Island, the inspector had asked his opinion on the resumption of a green turtle fishery similar to the fisheries in the 1920s and 1930s for the soup canneries. He told the inspector it could not be resumed, and he recommended instead a program of total protection of green turtles and the other five species throughout the entire Queensland coast and waters. This would ensure the safeguarding of an important attraction for the growing tourism on the Great Barrier Reef. Bustard seemed to think that Queensland was about to enact the legislation solely on the strength of his recommendation. It had done so in the case of Frank Moorhouse, he said; and he, Bustard, had even better field data to support his recommendation.[32]

The meeting had further convinced Carr that it was essential to have a full-time turtle man at the Survival Service Commission, someone who had both a sense of urgency and the diplomatic skill to work with Wolfgang Burhenne, chair of IUCN's Commission on Legislation, on international agreements. Burhenne was anxious to know if the members of the working meeting had been able to pull together enough data on the status of marine turtles to support international and national laws for their protection. After the meeting Holloway had to tell Burhenne that they had not. The delegates had identified significant gaps in the data, which they proposed to fill with additional fieldwork. But population assessments took time, and it would be a few years before they had evidence of a worldwide decline sufficient to support an international ban on trade in turtles. Only one delegate, Mr. de Silva from Sabah, had expressed a desire to see modification of his country's legislation.[33]

Leo Brongersma did not let the matter of the executive officer rest. After the meeting, he wrote to Holloway with copies to all the delegates. He suggested a number of actions for the new Marine Turtle Specialist Group but reiterated his displeasure with the manner in which the final agenda item was handled.

Colin Holloway defended his handling of the matter by explaining to the delegates that he had agreed after the meeting to circulate a draft job description for the new position. No one he had spoken to had objected to Pritchard's appointment. Although he felt no need to apologize for his handling of the matter, Holloway tried to smooth the ruffled feathers. To give the new marine turtle group a good start, he wrote to the delegates that "the man recommended for the Executive Officer post (his appointment is dependent upon availability of funds) does not require 'a sporting chance', he needs the whole-hearted support and co-operation of every member of the Group. How else can he be expected to operate? Bigger ventures than this one have begun from a more shaky start. I would like to recommend to all members of the Group that more effort be concentrated on the start and less on the shaking."[34]

FIGURE 23 Working Group Meeting of Marine Turtle Specialists, Morges, Switzerland, March 1969. First row, from left to right: John R. Hendrickson, Colin Holloway, Leslie Hendrickson, and Stanley de Silva. Second row: E. Balasingham (in overcoat with turned-up collar), Antonio Montoya (in light suit), Archie Carr, and Harold Hirth. Third row: Leo Brongersma (arms crossed, in glasses and dark suit), Dietrich Sahrhage, Mary Margaret Goodwin, and Tom Harrisson. Top two rows: Joop Schulz (in glasses, partially hidden), Mr. and Mrs. Webb, Robert Bustard (in dark overcoat and scarf), George Hughes (behind Bustard), Peter Pritchard, and Barbara Harrisson (peering out behind Tom Harrisson).
Courtesy of Peter C. H. Pritchard.

FIGURE 24 Archie Carr, Tom Harrisson, and Peter Pritchard making plans for Marine Turtle Specialist Group at Pritchard home, Orlando, Florida, early 1970s. Courtesy of Peter C. H. Pritchard.

When the Treaty Collapses

Archie Carr had told the delegates at the meeting in Morges that Costa Rican officials seemed receptive to the idea of establishing a national park at Tortuguero. Even though nesting numbers were in decline, if the government took this action now, the park would completely protect the green turtle nesting colony, closing the loophole that allowed turtles to be taken just offshore and the calipee poachers to disguise their illegal harvest. But when he got home from the meeting, he heard that Costa Rican officials had abandoned the park idea when they learned of the construction of another turtle freezing plant in Nicaragua. The first plant opened in late 1968, and Barney Nietschmann told him it was buying large numbers of turtles from the Miskito, bribing them with easy credit, cheap nets, and a ready cash market for the turtle catches.

Carr wrote straightaway to the minister of agriculture and ranching, Guillermo Yglesias, urging him to reconsider. He pointed out as tactfully as he could that if Costa Rica allowed the slaughter of Tortuguero nesters to continue, it would be much more damaging to the species than the Nicaraguan factories. Tortuguero was the green turtle's only nesting beach in the western Caribbean; Nicaragua's waters were only one of the colony's feeding grounds. For the colony to survive, it had to be completely protected at the nesting beach. He urged Minister Yglesias instead to negotiate with Nicaragua to reduce the take of turtles so that the population could be restored "as an important food resource for Caribbean people." At Carr's urging, Peter Scott sent a very similar letter to Yglesias, as chairman of the Survival Service Commission.[35]

But Carr did not stop with letters. He supported the efforts of the Caribbean Conservation Corporation's Billy Cruz to bring the fisheries ministers of Costa Rica, Nicaragua, and Panama to a meeting in San José that September. His tag returns showed that in fact these three countries held the key to the green turtle's survival in the western Caribbean. At least two-thirds of the nesting colony in Costa Rica foraged in the seagrass pastures off the Miskito coast of Nicaragua.

Carr was delighted when he got news of the meeting from Cruz. After three days of discussion and deliberation, the ministers agreed to a three-year moratorium on all green turtle harvesting in their waters. In the interim, they would work out a regional plan for a carefully controlled harvest that would not drain the nesting colony in Costa Rica but would give Costa Rica a fair share of the allowable catch.[36] Carr did not attend, but Billy Cruz was there, as was Robert W. Ellis, the specialist from the FAO who had accompanied Robert Schroeder on his visit to Nicaragua's fisheries development agency earlier in the year. Carr telephoned everyone he could think of with the good news. They now had an international but locally grown agreement that was fully responsive to the tagging study's results. He contacted Scott and Holloway at IUCN headquarters suggesting that they congratulate the ministers on their achievement.[37]

When Carr heard what the ministers did with the agreement after returning to their capitals to seek ratification, he was crestfallen. The official from Nicaragua's INFONAC reneged, deciding the freezer plants should put even more effort into catching and processing all the green turtles that could be taken. When Costa Rica heard that Nicaragua intended to keep sending turtles hatched on Tortuguero to the slaughterhouses, it too repudiated the agreement. Yglesias allowed the two turtle processing factories in Limón to keep buying turtles from the boats that cruised along the coast off Tortuguero. These boats harpooned mating pairs or females going ashore or swimming away from the beach after nesting. The harpooners made it futile to protect the nesters on the beach. The national park idea was dead.

What Carr found especially dismaying was that the governments were doing this with full knowledge of the risks. It was no longer 1954; more was known about the western Caribbean green turtle's life cycle and vulnerability to overexploitation than about any other population. They also knew that the decline was already in progress. They justified encouraging an export market on the need for economic development in the poor coastal regions. But Carr knew the exploitation rates this market was driving would obliterate the species—plain and simple. And the local people would no doubt suffer. Nietschmann had learned that in his research in Tasbapauni, the Miskito village that was now selling every last turtle to the company.[38] Carr wrote: "So the world's first international agreement for sea turtle conservation has disintegrated, and about all we can do is wait and see whether new losses, sure to come, will scare the decision-makers into making a new effort at cooperative control."[39]

CHAPTER TEN

Reptiles on the Red List

When the March 1969 meeting in Morges was over, the members of the sea turtle specialists group dispersed to their respective regions and elsewhere, to assemble more data on marine turtle populations of the world and on the global trade that now threatened the green turtle's survival. Archie Carr went right back to work on the migration problem. He caught a military transport flight to Ascension Island in the South Atlantic, where the green turtle nesting season was in progress. There, he put telemetry gear on three adult turtles and released them 100 miles offshore in order to track their course back to the beach.[1]

Tom Harrisson returned to Ithaca, New York, where he was now associated with Cornell University's Southeast Asia Program and teaching in the anthropology department, and where his wife, Barbara, was pursuing a PhD. Before turning back to his various book projects on Borneo, he wrote up an account of the meeting in Morges and sent it to his friends at the Fauna Preservation Society, who published the journal *Oryx*.[2] The specialists attending the meeting had agreed to wait until global statistics on the turtle trade could be compiled before mounting a full-scale depopularization campaign against gourmet turtle soup and luxury products. But Harrisson wanted to keep green turtles in the minds of conservationists before the growing environmental movement got too focused on pollution, supertankers, and oil spills.

Harrisson was not sure how much longer he would stay in the United States; he was not enamored of Americans, and he felt the need to restore his name among his contacts in academia. His reputation had suffered some damage when some disgruntled archaeologists he had tangled with spread rumors that he had taken hundreds of artifacts belonging to the Sarawak Museum when he left Borneo in 1967.[3] This was not true, but he had offended so many people in his various endeavors on behalf of Borneo's natural history that his enemies were likely to believe he was capable of theft. He was also interested in following up on a flirtation that began while he was in Morges. At an afternoon reception, he was flattered by the attentions of a wealthy woman who was presenting a sculpture to IUCN. Baroness Christine Fornari lived in Brussels, a city that was quite a bit closer to the kind of international activities Harrisson was eager to stay active in. He thought perhaps he could revive his social survey work from the Mass Observation days on behalf of the green turtle and other vanishing species that were victims of foolish fashions, and spend more time in Europe in the process.[4]

Peter Pritchard contacted all the turtle specialists and asked them for materials so that he could compile a synthesis of all available information on the leatherback turtle; he also asked them to return a questionnaire that would be used for the next Red Data book on reptiles. Following up on Hal Coolidge's idea, John Hendrickson made travel plans to survey potential sites for sea turtle farms in the South Pacific for the FAO, joining forces with Hal Hirth for the trip. Hirth had recently completed a study for the FAO on the sea turtles of the Gulf of Aden. At Carr's suggestion, Hirth offered to write a synopsis of the sea turtles for the FAO's series on food fishes.[5] Carr wanted to keep the UN agency's attention on the value of sea turtle fisheries and on the need to manage these fisheries, if they were to continue, and perhaps to invest in turtle cultivation.

From the Morges conference Robert Bustard went to his home in Perthshire, Scotland, and then stopped in Fiji on his way back to Australia. There, he found strong conservation laws in place but very few green turtles.[6] Poaching was rampant, and government enforcement resources thin, as local people hunted green turtles for food. If this was happening in Fiji, it could happen in northern Queensland, where an aboriginal fishing exemption applied to turtles and dugongs. When Bustard got back to Canberra, he set out to find government funds with which to bring cottage-scale turtle farms to the indigenous communities in the Torres Strait Islands. He also considered starting an experimental crocodile farm where there were still a few saltwater crocodiles left in northern Australia, on the aboriginal lands along the Edwards River. Realizing the value that regular meetings with crocodile experts would have toward such efforts, Bustard decided to ask whether IUCN's Survival Service Commission was interested in sponsoring a crocodile specialists group.

In July 1969 the Survival Service Commission formally approved the membership of the new Marine Turtle Specialist Group at the request of Colin Holloway. At the same time, it approved Holloway's draft description of the group's mandate, which stated that the group would "initiate action to re-establish the marine turtles as a viable resource throughout their natural ranges," as well as compile data for the Red Data book.[7] When René Honegger, the compiler of the Red Data Book for amphibians and reptiles, circulated an updated version of the marine turtle pages for the 1970 revision of the book, he asked the specialists to check the status categories and see if they met the latest version of the IUCN's classification criteria. The new status categories and their criteria were as follows:

1. Endangered: in immediate danger of extinction and continued survival unlikely without implementation of special protective measures;
2. Rare: not under immediate threat of extinction but could quickly disappear and requiring very careful watching;
3. Depleted: still occurring in numbers adequate for survival but heavily depleted and continuing to decline.[8]

Bustard did not wait to receive Honegger's letter before suggesting that the green turtle be reclassified; he wrote to Honegger as soon as he heard that the IUCN had adopted new classification criteria. He wanted to be sure the 1970 data sheets reflected the good conditions that prevailed for green turtles in Australia. At Bustard's suggestion Honegger added language to the green turtle data sheet noting that turtles were protected "throughout the State of Queensland over a coastline of 3200 miles and over 1200 miles of the Barrier Reef. In 1968 the Northern Territory of Australia banned commercial exploitation of the green turtle. In Western Australia turtles may only be taken by the licensees fishing in clearly defined areas. Should any signs of overfishing appear these licenses will be revoked." In the section on population status, Bustard had suggested adding: "There are sizable populations nesting on cays in the southern part of the Barrier Reef where they are a great tourism attraction. On Heron Island alone about 250,000 eggs are laid in a good season. . . . This species is also abundant in northern Australian waters, including the islands of the Torres Straits." In his letter to the specialists asking for further suggestions, Honegger noted in a postscript that Bustard had recommended that the green turtle be status category 3 (depleted) rather than 1 (endangered), "looking at the situation on a worldwide basis."[9] Bustard's view did not prevail, however, and the data sheets for the 1970 Red Data Book on reptiles assigned the same status categories to marine turtles as had the 1968 book.

As far as the IUCN was concerned, in 1970 the green turtle was still endangered. Controls on beach development, a reserve at Aldabra atoll, and experimental turtle farming were still the principal protective measures proposed. The new data sheet retained the language Carr supplied for the 1968 data sheet, that "there seems no inherent reason why *Chelonia* should not become a semi-domesticated meat animal of great value. Successful evolution of such culture should not only extend the means of taking food from the sea but would quickly take the pressure off of wild sea turtle populations, and thus help save the species for the distant future."[10]

The Secretary's Lists

Back in 1967, when Archie Carr was working on the first Red Data Book sheets for sea turtles, he hoped that the US Department of Interior would add reptiles to the informal list of species it believed were threatened with extinction. He had made the best case he could for their protection in the 1963 book *The Reptiles*, which he wrote for the Time-Life Nature Library series. At IUCN's conference in Nairobi in 1963, he had heard Stewart Udall, the secretary of interior in the Kennedy administration, make some encouraging remarks and was impressed with the secretary's commitment to conservation.[11]

Attending the IUCN conference in Nairobi had reminded Udall of how little his department had been able to do to stem the loss of native species in the United States. After the US Congress enacted the Land and Water Conservation Fund Act in 1964, his department was authorized to use monies from recreational user fees, a tax on outboard motor fuels, and other related sources to acquire habitat for native wildlife. Udall's Fish and Wildlife Service made plans to expend some of these funds to buy land for several vanishing bird species, including the whooping and sandhill cranes, the Mexican duck, and a number of endangered Hawaiian birds. But members of Congress with power over the department's budget questioned the wisdom of such large expenditures, given the dwindling numbers of individuals in these populations.[12] Language in the 1964 act specified that the fund could be used for "the preservation of species of fish and wildlife that are threatened with extinction." The directors of Fish and Wildlife heard from these members of Congress that the bureau would need new federal legislation specifically authorizing purchases of land for native species preservation.

Udall realized that his department would need a more authoritative list of species that required intervention for their survival. He directed his staff to set up a committee on rare and endangered wildlife species, modeled on IUCN's Survival Service Commission, to advise the department on which species of native vertebrates were threatened with extinction. By July 1964, the committee, composed of nine biologists from the Bureau of Sport Fisheries and Wildlife and refuge managers, circulated a preliminary list drawn from the recommendations of more than 300 species specialists. The final list in August—called the "Redbook"— contained sixty species and subspecies of vertebrates, including sixteen mammals, thirty-five bird species, six fish, and three reptiles. The department also made plans to create a captive breeding facility in Patuxent, Maryland, to follow up on its recent success in breeding whooping cranes.[13]

Udall's staff also set about drafting the necessary legislation to give the department authority for a comprehensive species preservation program. It was clear that the department would need the power not only to purchase habitat that the species depended upon, but also to propagate selected species of native fish and wildlife that had been extirpated by hunting and land development.[14] Some members of Congress questioned the need for the legislation in view of the fact that wildlife preservation was generally regarded as a state law function. Nevertheless, both the House of Representatives and the Senate passed the legislation without significant controversy. Among its features, the law required Udall's Interior Department to continue to compile and maintain an official list of native vertebrate species. The new law did not dictate preservation policies to the states but concentrated instead on directing the federal departments and agencies to protect the species on the list wherever it was both "practicable" and consistent with their missions.

The 1966 Endangered Species Preservation Act was not quite the comprehensive authority that Secretary Udall had envisioned, but it was a start. The department quickly published the list of the now seventy-eight species that, according to its information, were at risk of extinction. In addition to the list, the new law authorized up to five million dollars for the department to spend acquiring wildlife refuges to protect these species. But nothing in the bill authorized the department to stem the land development and market hunting outside the wildlife refuges that were driving numerous native vertebrates ever closer to extinction. When the ink was not quite dry, Udall directed his staff to draft a bill to enhance the 1966 act with the authority that was really needed.[15]

The deficiencies of the 1966 act quickly became apparent with respect to the American alligator (*Alligator mississippiensis*).[16] A native species of southeastern North America, whose plight Archie Carr had written about with such eloquence in *National Geographic*, the alligator had fallen victim to the social fashion of wearing exotic bird feathers and animal skins that grew in the nineteenth century along with the taste for green turtle soup. As the native green turtle was stripped from Florida's keys and beaches and shipped live and in cans by rail and steamboat to the north and across the Atlantic, the alligator was supplying shoes, bags, and tourist curios to the growing numbers of people who visited Florida.[17]

The US Fish Commission had been aware of the alligator's poor prospects at the turn of the century but had not taken steps either to manage the hunting pressure or to cultivate a stock to replenish the wild populations. Hugh Smith, the same federal ichthyologist who had read Ralph Munroe's plea on behalf of the green turtle at the 1890 Fishery Congress in Tampa, wrote a report for the US Fish Commission describing the scope of the alligator trade. With 2.5 million alligator hides shipped out of Florida in the previous dozen years, Smith warned Congress that the alligator could not support this level of commerce. In Florida, Smith wrote, the alligator was being "systematically and relentlessly hunted" in virtually every corner of the state; it was only a matter of time, he said, before "this valuable fishery resource will become exhausted."[18] But there was no Ralph Munroe to quote on the need for intervention to protect this native animal; no citizens of Florida or anywhere else had taken up the alligator's cause as they had for the American bison and the birds with fancy plumage.[19] The alligator was a fearsome menace in the popular view, and if it went the way of the passenger pigeon, then at least the fashion had been diverting.

It was not until the 1920s that Southern nature lovers came to the alligator's rescue. Louisiana and Florida adopted limited hunting seasons—but exempted the parishes and counties where most of it took place. Then in 1943, Florida enacted a closed season during alligator breeding months and a year-round four-foot minimum size, but this did not stem the decline. The alligators' swamps were being drained for agriculture and housing to support Florida's burgeoning population, and hunters were able to penetrate the remaining swamps with airboats and swamp buggies.[20] By the time the Interior Department added the alligator to its first Red Book, the animal had a number of champions. But these citizens and game wardens knew that illegal hunting was still rampant, even after Florida banned all alligator hunting in 1961, and that the alligator and its habitat were fast disappearing.

In his 1967 article in *National Geographic*, "Alligators: Dragons in Distress," Archie Carr argued that it was not enough to ban hunting; these laws could be evaded simply by crossing state lines into places that lacked the restrictions. To stop the slaughter it was necessary to end the fashion for alligator hides. It was also essential to extend the Lacey Act of 1900 to reptiles. That law had been indispensable in ending the plumage hunting in Florida by making it a federal offense to cross state lines with wildlife taken in violation of state game and wildlife laws.

Udall's staff wrote a new version of the legislation to extend the Lacey Act to reptiles. The bill sailed through the House of Representatives but hit a snag in the Senate. The fur, leather, and pet industry mounted a campaign to weaken the bill, arguing that species extinction was an international problem that the United States could not successfully tackle on its own. After extensive negotiations with the industry lobbyists, Interior staff resubmitted the bills. Finally, in late 1969, Congress passed a revised version that allowed the department to list species of reptiles, amphibians, and invertebrates but which included several concessions to the wildlife industry. The new president, Richard M. Nixon, signed it into law on December 3, 1969.[21] The Lacey Act was expanded to cover all wildlife taxa, including amphibians, reptiles, mollusks, and crustacea, as Carr had sought.

The fur industry, however, had succeeded in limiting the department to listing only those species that were threatened with "worldwide extinction." This standard would require more data than the department had in many instances and prevent the department from listing until a species was already at critically low numbers. Fur manufacturers had also insisted that the act create a level playing field: if they were going to be prohibited from importing skins, they wanted their foreign competitors to be barred from exporting skins to the United States. The 1969 law gave the Interior Department the power to ban both interstate and foreign commerce in endangered species. It further required the department to convene a conference to negotiate an international treaty so that all countries would agree to prohibit trade in the same imperiled wildlife species.

Archie Carr was hopeful when he saw that the new law directed the Interior Department to convene a conference where nations could negotiate a treaty to control international trade in endangered species. This action would dovetail nicely with the interest of IUCN's Survival Service Commission in such an agreement and his own desire to see an international agreement among the Caribbean nations with sea turtles. Carr had confidence that Wayne King, the director of conservation for the New York Zoological Society, who had worked hard on behalf of the 1969 legislation, would make sure there were as few loopholes as possible in the new wildlife trade treaty. King had been Carr's research assistant at Tortuguero one summer in the early 1960s, was now his comrade-in-arms in the campaign to save the American alligator, and would soon champion the Atlantic green turtle.

Interior's Fish and Wildlife Service wasted little time in preparing to implement the new law, especially the provisions on permits for importation. The law was due to take effect in early June 1970. In April, Fish and Wildlife published a proposed list of endangered species along with proposed regulations and invited the public to submit comments, including objections or suggestions. On the proposed list were a total of ten reptiles, including three species of sea turtles and three species of Crocodylia. The three sea turtles were the green turtle (*Chelonia mydas*), the hawksbill turtle (*Eretmochelys imbricata*), and the leathery turtle (*Dermochelys coriacea*).[22] But when the final rules were published in June, the green turtle had been dropped from the list. A notation to the final list said only that "if a candidate species is not listed it may be because it is not endangered throughout its range or because there is insufficient evidence to warrant its inclusion on the list at this time. The list is under continual review. Factual data are welcome and should be submitted."[23]

Two months later another notice appeared, explaining that when the final list was published, the Fish and Wildlife Service had deleted all species that had not previously appeared in a notice of proposed rulemaking. The Interior Department was now proposing most of those same species and adding additional species. The list now included the Atlantic ridley (*Lepidochelys kempii*), which was referred to by its common name, the Kemp's ridley. When the final list was published in December, the green turtle was still missing from the list. It was now official; in the opinion of the Department of the Interior, the green turtle was not threatened with worldwide extinction, but the hawksbill, the leatherback, and the Kemp's ridley turtles were.[24]

Adventures in the Skin Trade

After Wayne King finished his master's degree in biology at the University of Florida, he had headed off to the Field Museum in Chicago to work on a PhD with Robert F. Inger on his National Science Foundation–funded study of rainforest frogs. Inger had collected reptiles for the Sarawak Museum in the 1950s and had become a friend of Tom and Barbara Harrisson. He sent his new student to Borneo to begin the research. King spent nearly a year in the Sarawak lowland rainforests, working with local Iban tribesmen to collect seventy-two species of reptiles and amphibians and measure the rainfall.[25] While in Sarawak, King got to know Tom Harrisson, who was overseeing an archaeological dig at the Niah Great Cave, and Harrisson asked King to identify the animal bones that were found in the cave.

When King and Inger had a disagreement about the direction of the Field Museum's frog project, King left the project and transferred to the University of Miami. After conducting a study of lizard species introduced in Florida and earning his PhD, King considered returning to Borneo to study the flying lizard and work with Tom Harrisson on Sarawak Museum's reptile collection. But by the mid-1960s, the political atmosphere in Borneo was tense, and Harrisson warned him he would not be safe collecting specimens in the rainforest. King went instead back to Gaines-ville and the Florida Natural History Museum. He watched the politics of conservation as Florida was pressed to enact tougher laws to protect the American alligator and its habitat. After a brief teaching stint in California, King found a position as the assistant curator of reptiles for the New York Zoological Society and set about campaigning for protection of all the reptiles and mammals that were disappearing to the exotic skin and fur industry.

The New York Zoological Society, founded in 1895 by several prominent New Yorkers, including Theodore Roosevelt, had a long history of species conservation, beginning with the work of William Hornaday. As director of the New York Zoological Park, Hornaday was instrumental in saving the Pribilof fur seals and the American bison from extinction and campaigned tirelessly for wildlife conservation laws at the turn of the century.[26] In recognition of his successes, the founders of the society endowed a trust fund to support curators in such work. When King showed interest in seeing these funds used to research vanishing species, his boss at the zoological society, William Conway, asked King to serve as the society's director of conservation.[27]

Throughout the 1960s, the fashion of wearing coats made from the skins of the spotted great cats—leopards, cheetahs, jaguars, and the like—took off, and wildlife biologists and game wardens around the world reported sharp drops in these populations. The Interior Department now had a small Office of Endangered Species and International Activities to determine which species were threatened with worldwide extinction. When this office balked at listing the great cats because they were used commercially, King realized that the 1969 Endangered Species Act would not be able to stem the drain on the caimans and crocodiles throughout the tropics that were being hunted so heavily for their skins. He could see as well that the exotic leather trade had begun to drain the Pacific ridley turtle populations when companies in Mexico discovered large nesting populations along the Pacific coast. When he mentioned this to one of the Zoological Society's trustees, the trustee asked William Conway to see what could be done.

In the spirit of Hornaday, Conway and King hired a private investigator to determine whether the major shoe and handbag manufacturers were obtaining their leather from illegal hunting and whether they knew the trade was illegal. The detective found that most of the illegal skins went to tanneries in New York. King worked with the leather goods designers and manufacturers to help them develop a conservation ethic; and he and a colleague, Peter Brazaitis, published a paper to assist the tanneries in identifying the species used in commercial reptile skins.[28] King convinced Claire Hagen, who designed purses and shoes for the designer Bill Blass, to find out where her turtle skins were coming from. She flew to Mexico and told the three tanneries she worked with to stop sending her turtle skins and focus instead on finding the softest calfskin.[29]

King knew it was not enough to change the views of one designer in one location; others would simply buy the leather, and the drain would continue until every caiman and sea turtle was gone. It was necessary to have a global approach. At Conway's suggestion, he began attending IUCN meetings, particularly those of the Survival Service Commission, rising to the rank as vice-chairman under Sir Peter Scott.[30] He spoke frequently with Carr's former student David Ehrenfeld, who was now teaching at Barnard College in New York City, and encouraged Carr to mobilize the commission's Marine Turtle Specialist Group into conservation action. He recommended that the commission establish a Crocodile Specialist Group to revise and expand the assessments for the Red Data Book volume on reptiles and amphibians.

King became even more active in the campaign to protect the American alligator in Florida. He drew upon the resources of the Zoological Society to co-sponsor a series of symposia that brought together alligator experts from many walks of life: academia (including Archie Carr), fish and game agencies, and nongovernmental organizations dedicated to wildlife. At these meetings, King could see that competing philosophies were animating the discussions. Some experts, like Carr, thought only a complete ban on hunting would prevent extinction. Others argued that the alligator could still be used as a resource if hunting and land conversion were managed more strictly. Despite these differences, the assembled all agreed that in addition to more research and public education on the plight of the alligator, it was necessary to pursue additional legislation at the local, state, and federal levels. By the end of their third symposium—which, like the turtle experts' meeting in Morges, took place in March

1969—the alligator experts adopted the name American Alligator Council, patterned somewhat after Carr's successful Caribbean Conservation Corporation.[31]

Because the United States was the biggest market for exotic skins, King assumed responsibility for the American Alligator Council's legislative campaign. New York City, with its hundreds of fashion designers and clothing manufacturers, was the center of the US market. He lobbied state legislators in Florida to get them to enact laws that would ban the sale of boots, jewelry, and other things made from alligator and that would send poachers to jail for violating hunting laws.[32] With the help of Mayor John Lindsay, King lobbied councilors for New York City for an ordinance to ban the sale of products made from alligator.[33] But his biggest victory by far was the Mason Act, an amendment to New York State's agriculture and marketing law.

King approached a member of the New York State Assembly and explained to him that the 1969 Endangered Species Conservation Act was failing the American alligator and the leopard. Skin dealers were evading the law by labeling their products as the skin of a species that was abundant but similar in appearance to an endangered species and could not readily be told apart from the one on the list. The Interior Department could have closed this loophole by listing species that closely resembled the endangered ones, but it had not. The Senate Commerce Committee had instructed Interior to list only those species and subspecies that were actually threatened with worldwide extinction.[34] While conservation groups were working on getting a "similarity of appearance" provision added to the federal law, it was crucial for the states to adopt laws prohibiting the sale of every species and subspecies of cat and crocodile that could possibly be imported.

Edwyn E. Mason was a member of the New York Assembly who was sympathetic to King's cause. As chair of the agriculture committee, he introduced a bill to amend New York's agriculture and marketing laws to prohibit the sale or possession for sale of all species and subspecies of the order *Crocodylia*, and of nine species of cats, the red wolf, the polar bear, and the vicuña, listing each one by its scientific name to prevent deliberate confusion of customs inspectors. And there would be no exception for the supposedly abundant subspecies that dealers had claimed to be importing. Mason convinced forty-four other Assembly members to co-sponsor the bill and then shepherded it through a contentious hearing in February 1970 to passage.[35]

It did not take long for New York's fur and skin industry to react. In a challenge to the Mason Act in state court, a leading manufacturer of crocodile shoes, joined by several fur-coat makers and retailers, denied that they used endangered species in their products. In any event, they argued, the Mason Act was unconstitutional. The trial court agreed; the act was too broad in not making a distinction between endangered and non-endangered species of cats and crocodiles, and was unnecessary and "unconstitutionally oppressive" in affording protection to species that were not native to New York.[36]

The state attorney general then mounted a vigorous appeal. In *Nettleson v. Diamond*, the state court of appeals affirmed the Mason Act's constitutionality, finding it was not unreasonable for New York's legislature to enact such a broad ban if that was the only way to protect species that were imperiled by commerce. Relying on the expert opinion of Dr. F. Wayne King that processed hides of endangered crocodile species cannot be distinguished from the hides of abundant species, the court found that the broad ban was necessary.[37]

A Massachusetts company had filed a similar case in the federal district court in Manhattan, but the federal court stayed the case to let the state court rule on the validity of the Mason Act. As soon as the Nettleson appeal was decided, the federal district judge, Judge Mansfield, concluded the state court was correct. New York indeed had the power to legislate in order to prevent the extinction of distant species. Judge Mansfield described the importance of the state's interest:

It is now generally recognized that the destruction or disturbance of vital life cycles or of the balance of a species of wildlife, even though initiated in one part of the world, may have profound effect upon the health and welfare of people in other distant parts. We have come to appreciate the interdependence of different forms of life. We realize that by killing certain species in one area we may sound our own death knell. . . . Nowhere does the Secretary of Interior indicate that his list of endangered species is definitive. The state's . . . list may be broader than the federal list simply because the state legislature did not see fit to wait until only a handful of species remained before it passed a law affording protection. We cannot overrule the legislature for being cautious. Extinct animals, like lost time, can never be brought back. They are gone forever.[38]

Not willing to throw in the towel, the Reptile Products Association asked the US Supreme Court to review *Nettleson v. Diamond*, the New York state court decision that federal district judge Mansfield had found so persuasive. On March 23, 1971, the Supreme Court, declined to hear the appeal, and the Mason Act went into effect. Wayne King wasted no time in making the same case to state legislators around the country that he had made to New York Assemblyman Mason. The difference was that now, King could add that the New York law had been upheld by the highest court in the land. In short order, California enacted a similar law; and four other states, including Massachusetts, the home of many crocodile shoemakers, began to consider adopting their own endangered species laws. But with so many state laws banning imports of crocodile skins, manufacturers began to import sea turtle skins.[39]

CHAPTER ELEVEN

You Lost the Turtle Boat

On March 23, 1971, the day the US Supreme Court let the Mason Act's ban on exotic skins take effect in New York, the merchant vessel *A. M. Adams* moved slowly up to the wharf at the foot of Margaret Street in Key West, Florida. Built on Grand Cayman Island for Norberg Thompson's green turtle cannery in Key West, the *Adams* had been the largest and fastest of the Caymanian turtle schooners. Skippered in her heyday by Captain Allie Ebanks, she was now a motor vessel, her back deck covered with a chunky deckhouse and her masts shortened to hoist cargo instead of sail. But her cargo was the same as it had been for three decades: giant green turtles lying supine on every available inch of the deck and in the hold. She was the run boat now; her job was to carry to Key West the green turtles caught near the Miskito Cays off Nicaragua by the four remaining catching boats from Grand Cayman. The captain reported that the turtles were caught "around the Cayman Islands," but everyone knew they were taken illegally from the Miskito Bank.[1] That morning, the *Adams* was bringing only 135 turtles, a far cry from the 400 to 500 turtles the *Adams* was accustomed to hauling.

The crew could see the wharf was crowded with the usual cast of characters: the men from Sea Farms, Inc., which now owned Thompson's former cannery, and a small cluster of tourists waiting to see the giants hoisted onto the dock and slid down into the kraals. The crew also recognized the reporter from the local newspaper and thought that perhaps Bill Hannum, the president of Sea Farms, had organized a publicity stunt. These days the company's main source of revenue was the tourists who ate at the restaurant and paid admissions to see turtles swimming in the kraals. But why was the bureau chief from the *Miami Herald* on the dock, with his camera slung around his neck?

As the eight-man crew began to hoist the turtles onto the dock, they heard someone shout for them to stop. It was Don Sweat, the marine biologist who worked for Sea Farms. He told them to keep the turtles suspended from the hoisting arm until he could measure the length of their shells. If they were less than forty-one inches long, Hannum had told him, Sea Farms could not keep them in the kraals or sell them up north. Making a great show of measuring each turtle for the reporter's

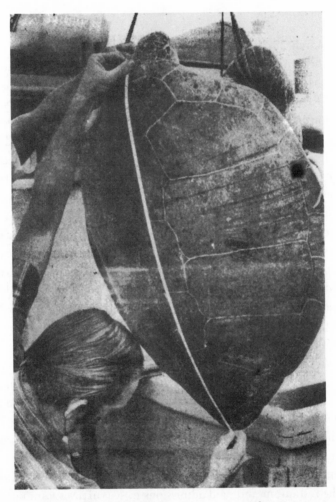

FIGURE 25 Turtle cannery biologist measuring *A. M. Adams*'s green turtle landings for compliance with Florida's new forty-one-inch minimum size regulation, Key West.
From "Turtle Industry Faces Extinction," *Miami Herald* Mar. 29, 1971. © The Miami Herald, 1971. Photograph by Wright Langley.

camera, Sweat shouted out each measurement. When he was done, only five were forty-one inches long. The average length was thirty-five inches.

Bill Hannum later told the reporter that the trip was a bust, and the crew might not get paid. But, he said, "don't blame me, blame Governor Reubin Askew."[2] The week before, the governor and his cabinet, miles up north in Tallahassee, had set a forty-one-inch minimum size rule for green turtles in Florida. The governor had set "an unrealistic size limit on turtles," Hannum said. "You can't find them that size, so we will cease operations, right now—we never made a nickel on the operation, anyway."[3] Hannum also blamed the citizens of Key West and the local newspapers for not taking his warning seriously. And what about the turtle fishermen? "I sent all the men home yesterday and that's the end of it, as far as I'm concerned. You lost the turtle boat."[4] Twenty-seven cannery workers were out of a job, and the *A. M. Adams* was tied to dock, awaiting her conversion to a snapper boat.

Bob Ingle had taken Hannum's warning seriously, however. He was the director of the marine science bureau of Florida's Department of Natural Resources. The week before the arrival of the *Adams*, he had proposed a less restrictive minimum size regulation to Governor Askew and his cabinet. He recommended a minimum size of twenty-six inches for green turtles caught off Florida's Gulf coast and thirty-one inches for those off Florida's Atlantic coast, including the Florida Keys. "My sizes were a compromise, but in the interest of conservation and protecting the industry they are reasonable," he told the *Miami Herald* reporter. But, in his opinion, the cabinet was too eager to stay on good terms with the environmental lobby and rejected his proposal in favor of the larger, forty-one inch minimum size on both coasts.[5] And this was the result; the "turtle boat" was out of business.

Florida's Failing Turtle Fishery

Turtle processing was no longer a big industry in Key West, but a decade earlier the arrival of the turtle boats had been a major tourist attraction, a "must-see" in every traveler's guide to the Keys. In the early to mid-1960s, turtle schooners regularly unloaded hundreds of green turtles, some as large as 500 pounds. In 1962, a schooner would arrive and unload its cargo every four to six weeks. The docks often swarmed with visitors, families with children and onlookers from the nearby docks. By the late 1960s, however, six months would pass before a schooner arrived with a hold full of turtles. By 1970 the turtle boat arrived only once or twice a year. Ingle knew the decline in green turtle landings was because the turtles were from the dwindling population on the Miskito Bank.

Over time the Nicaraguan government had tired of the Caymanians' fishery and made other plans for its green turtles. First, it raised the taxes per head of turtle and then, in 1965, established a closed season from mid-May to mid-July. Finally, the government simply failed to renew the treaty that allowed British-flag vessels to fish in Nicaragua's waters. Now, the Miskito fishermen from Tasbapauni and other villages, whom Barney Nietschmann lived among and studied, caught as many turtles as they could and sold them to the three Nicaraguan turtle processing factories, which froze the meat and carcasses and shipped them by the thousands directly to the markets in New York and Europe.

The *Adams'* crew knew all about the factories. The Miskito men they used to hire to tend their offshore kraals told the crew that they now only sold their turtles to the boat that worked for the turtle factories. When the company boat visited them out on the cays, the captain bought every last turtle, even the ones the men meant to bring home to their villages. The men on the *Adams* knew that this would be its last trip as the turtle boat of Key West.

Bob Ingle had felt differently about green turtle regulations when he was a young man. In 1949, he and his mentor, Professor Frank G. Walton Smith of the University of Miami, prepared a report on the turtle industries of Florida, the Gulf of Mexico, and the West Indies, complete with an annotated bibliography of every paper he could find on sea turtles.[6] It was clear to him then that Florida's green turtle fisheries were in dire need of management. The report recommended a complete ban on the taking of eggs and of nesting females, a minimum-size regulation to ensure that turtles reached sexual maturity before capture, and a hatchery and restocking program. These were very similar to the US Fish Commission's recommendations to Congress a half century before. In 1953, the Florida legislature adopted only one of Ingle's recommendations. It enacted a law making it illegal to "molest" a green turtle's nest by removing its eggs, even though it was rare for anyone to see a green turtle nesting in Florida.[7]

By 1956, Ingle was the chief of research for Florida's Board of Conservation. To verify sporadic nesting reports, he asked his conservation agents to watch for any green turtle nestings. On July 11, 1957, T. C. Cheatham saw a female nesting two miles north of Vero Beach, in Indian River County. This was the first report of a green turtle nesting in Florida since the 1890s.[8] When a second nesting was reported in July 1958 on Hutchinson Island, Ingle told the agent to carefully remove the eggs and rebury them in a makeshift hatchery at the nearby House of Refuge Museum, to implement another of his 1949 recommendations.

At the hatchery, a conservation board employee named Ross Witham watched over the nests. When the eggs hatched, Witham called Archie Carr, who immediately drove down and brought forty hatchlings back to his lab in Gainesville. Up to this point, Carr was uncertain whether Florida had in fact been a center of reproduction. He thought early writers like John James Audubon might have assumed so because Florida had large feeding aggregations of green turtles and nesting populations of loggerheads.[9] But the nests sighted in 1957 and 1958 convinced Carr that Florida once hosted great nesting assemblages of green turtles.

Once he knew that green turtles still nested in Florida, Carr decided to try to interest the state in restoring the population. He invited Bob Ingle to co-author a paper in the University of Miami's *Bulletin of Marine Science of the Gulf and Caribbean* describing the sightings. Ten years before, Ingle had started the oyster research and reseeding program for Apalachicola Bay and seemed to believe in taking positive action to restore depleted marine resources. In the paper, Carr connected the nest sightings to his NSF-funded migration studies, musing that whether the female turtles migrated as isolated individuals or in aggregations, they probably had trouble finding mates. He knew that green turtles mated sometime around nesting time but thought that the eggs a female laid in one season were probably fertilized three years earlier. His own observations at Tortuguero seemed to corroborate those of Tom Harrisson and John Hendrickson on Sarawak's Turtle Islands.[10] So in addition to their remarkable feats of navigation, female green turtles could probably store sperm and lay viable eggs three years later.[11]

Whether the turtles traveled alone or as part of a group, mated in that season or in the previous one, Carr now knew that female green turtles were arriving on Florida's beaches with fertilized eggs and ready to nest. It might be practicable, then, to devise a management program to restore Florida's lost colonies of green turtles. All that was needed was a nursery where green turtle hatchlings could be impounded in warm seawater and fed chopped-up fish and invertebrates until they reached six or eight inches in length. He wrote, "Although we still know nothing of the ecology of young sea turtles after they have passed the surfline, there can be no doubt that the releasing of a thousand half-pound turtles would be an operation more likely to increase the population than the natural emergence of a thousand hatchlings."[12] If volunteers systematically searching Florida's beaches found enough additional nesting females, the eggs from these nests could supply a pilot nursery. This was preferable to imported hatchlings because it was possible that the green turtle's ability to return to its natal beach was genetically programmed. But if only a few more nests were found, Carr thought bringing eggs or hatchlings from the Caribbean nesting beaches was worth a try. Green turtle hatchlings were very hardy and could be shipped in light crates or even bags without food or water and only a minimum of moisture.

As co-author of the paper that laid out this simple plan, Bob Ingle would be sure to endorse a restocking project, especially when the paper contained Carr's statement, "From every aspect, then, and at all stages of development, the green turtle appears to be a fit subject for management. It can only be a matter of time until somewhere this opportunity is exploited." But Ingle, ever the pragmatic manager, added a dose of reality in the last two sentences: "Although more immediately pressing problems preempt present conservation budgets in Florida it may be possible to undertake such a project in the future. The growing interest in the subject indicates that considerable public support could be expected."[13]

In 1957, the state legislature had taken baby steps in turtle conservation by amending Florida's nest protection law to prohibit the killing, taking, or possessing of any sea turtle "while such turtle is on the beaches of Florida or within one half mile seaward from the beaches" during the months of May through August. Anyone caught violating this law could be fined as much as $600 or sentenced to up to one year in prison. This change closed the loophole Carr had complained about that allowed fishermen to kill female turtles in the water before they had a chance to lay their eggs. Without it, the marine patrol agents had to catch the poacher in the act of robbing eggs from the nest.[14] Carr was initially pleased with these actions, but then it seemed that Florida took two steps back with every one step forward when it came to protecting green turtles. Shortly afterward, the Florida legislature amended its turtle law to allow an open season on taking sea turtle eggs in the small rural counties bordering the Gulf of Mexico and in St. Johns County bordering the Atlantic.[15]

Nevertheless, by the end of 1959, Carr was sending crates of green turtle hatchlings from Tortuguero to Florida's Department of Natural Resources for restocking. The new "head-start" program was headquartered at the House of Refuge Museum in Stuart, Florida, where eggs taken from the few nests laid on the east coast were incubating. Ingle had indeed taken the bait.[16]

The Green Turtle Is Not Endangered

After a decade of overseeing Florida's green turtle restoration program, Bob Ingle thought his state had the turtle problem well in hand. Then, in May 1970 Ingle heard that the US Department of Interior was adding green and loggerhead sea turtles to the list of endangered species. He was furious and fired off a letter of protest immediately:

> It is difficult to understand the method (or lack of one) by which [the proposed listing] was developed. Our information and experience with the two turtles does not support the contention *by anyone* that these species are endangered. While it may be true that the populations of these animals is not as great as in 1880, they are certainly far from a population density that might risk extinction. Although the Endangered Species Act concept is laudable, it loses effectiveness if species are carelessly and arbitrarily added to it without adequate justification.[17]

Whom in his research bureau had the Department of Interior consulted before making the proposal? Did the department ask anyone in the Southeastern Fisheries Association how the listing might affect its members who imported turtles and turtle products?

In reply, an Interior Department official, associate director J. P. Linduska, tried to smooth Ingle's ruffled feathers without conceding any error or misstep by his office. He explained that the department in fact had not proposed to list the loggerhead, only the green turtle, and that the lists were only initial lists, not the final ones. Interior had included the green turtle on the "candidate list of foreign species threatened with extinction" issued in March 1970, and then on "the tentative list of endangered species appended to the proposed Endangered Species Conservation Regulations published in the *Federal Register* on April 14." The department did not include the green turtle on the official list published subsequently in June because it had deferred the decision to list the green temporarily "to allow interested parties to submit data."[18]

Linduska hinted broadly to Ingle, however, that the green turtle would be listed eventually. He said that he understood this to be the official position of the state of Florida, having received on April 21 from H. E. Wallace, the assistant director of Florida's Game and Freshwater Fish Commission, a copy of a resolution on endangered wildlife that was about to be presented to the state legislature naming the green turtle as one of Florida's threatened species. Moreover, the green turtle had been on the Interior Department's red list since Secretary Udall's time. In 1966, the Fish and Wildlife Service had classified the green turtle, *Chelonia mydas mydas*, as "peripheral," that is, as a species "whose occurrence in the United States is at the edge of its natural range and which is rare or endangered within the United States although not in its range as a whole."[19]

Linduska told Ingle that when this classification was made, the Interior Department believed the green turtle was still abundant in the Hawaiian Islands. But the department had just learned that the state of Hawaii's Endangered Species Committee recommended redesignating the green turtle as endangered. For the species' status in Florida, his agency had relied upon the statement quoted on Interior's "RAP sheet" on green turtles which quoted from a publication by Carr and Ingle to the effect that it was "practically extirpated as a breeding entity in the fauna of the U.S.[20] He reminded Ingle that IUCN classified the green turtle as an endangered species because it "is so considered by many scientists including Dr. Archie Carr of Florida with whom I am sure you are familiar." His department had been told by conservation officials in other countries they were having a hard time controlling poaching due to the continued international trade. The new Endangered Species Conservation Act of 1969 required Interior to list species in danger of "worldwide extinction" and to prohibit their importation into the United States. Listing green turtles in the United States as endangered would assist these efforts.[21]

Ingle was not thrilled to have his own paper quoted back to him to contradict his current view that the green turtle was not endangered, which was the official view of his Department of Natural Resources. The constituents of Florida's Game and Freshwater Fish Commission were well-heeled sports fishermen, who were now in league with the conservation lobby and pushing for new environmental laws in the legislature. But commercial fishermen were the constituents of Ingle's department, and they worried that if sea turtles were listed as endangered, it would increase their cost of doing business. They commonly pulled up sea turtles in their nets and traps, dispatched them with a knife to the throat, and sold them to restaurants and fish houses for pocket change. Shrimp-fishing vessel crews, in particular, often butchered Atlantic ridley, loggerhead, and green turtles for their own consumption and for extra cash.[22] This had the added advantage of keeping turtles from fouling their gear repeatedly.[23] If the crew could not sell this accidental catch, boat owners would be under pressure to pay crew members a greater share of the boats' gross revenue.

Ingle doubted whether the Interior Department had any idea of these economic realities. Linduska told him that his staff had consulted the tanning industry, which used turtle skins, unaware that the Southeastern Fisheries Association had an interest in the endangered species list. For input on fisheries, the staff had consulted the Bureau of Commercial Fisheries, its sister agency in the Department of Interior.[24] The bureau's staff knew that shrimpers caught sea turtles but probably had not shared this fact with the endangered species office in Interior.

A Turtle in the Cabinet

The Florida chapter of the Izaak Walton League was the group of well-off sportsmen Ingle had in mind when he read Linduska's letter. In the late 1960s, the league was campaigning for laws to prevent land developers and the Army Corps of Engineers from destroying the marshes and estuaries that were the nursery areas for the state's rich wildlife. The president of the chapter was Bill Lund, an avid conservationist and treasurer of a new lobbying group, Conservation 70's, Inc.[25] Lund lived on

Jupiter Island, a barrier island on Florida's Atlantic coast, where, ever since boyhood, his son Frank had raised in their backyard the baby sea turtles he found while wandering on the island. When Frank went off to the University of Florida to study wildlife biology, he turned his hobby into the Atlantic Loggerhead Turtle Research Project, raising private monies to support a team of volunteer turtle watchers. Every season Frank and his crew patrolled the beaches and transplanted nests from the beach to a well-guarded hatchery to protect the eggs and hatchlings from predators. To avoid running afoul of the nest protection law, before each season he obtained permits from the Florida Board of Conservation.

In 1971 Frank Lund was in his junior year. When he met Archie Carr, who was a well-known professor at his university, and told him there were green turtles nesting on Jupiter Island, Carr encouraged him to enlarge the scope of his research to green turtles by counting and mapping their nests and measuring and tagging the females using the methods his research team used on the green turtles of Tortuguero.[26]

Lund divided the twelve miles of beach on Jupiter Island that extended south of St. Lucie's Inlet into one-eighth-mile segments and every year from June through early September recorded the location of every nest he found. Because the loss of eggs from predation and beach erosion seemed especially heavy on his beaches, Frank stepped up his nest transplantation and hatchery efforts and kept detailed records of the fate of the nests that remained buried on the beach. He carefully recorded the carapace lengths of the nesting females, noting a range in size from almost thirty-eight inches to over forty-four inches, with a mean of nearly forty-one inches.[27]

When Frank Lund saw that his mean length was almost two inches larger than the thirty-nine inches Carr had recorded for Tortuguero, he, like Carr, began to believe Florida had its own green turtle population. But he thought the turtles had a greater chance of survival if the nests were left in the beach and not moved to the restocking program. He decided to keep his nests a secret from Ross Witham and his head-start program.[28]

To help the sea turtle species that his son was so worried about, Bill Lund asked his local legislator to introduce a bill in the 1970 legislative session to require the Department of Natural Resources to maintain predator-free nesting preserves for all of Florida's sea turtles until they were no longer at risk of extinction. The bill also would require the department to implement a three-year moratorium on the taking of green turtles in Florida or its state waters. This would be Florida's contribution to the tripartite agreement on green turtles which, Frank had learned from Archie Carr, had been provisionally adopted by Costa Rica, Nicaragua, and Panama in September 1969.

The Lunds' local legislator happened to be the president of the state senate at the time. Senator Jerry Thomas saw the bill safely through the state senate, but the bill faltered in the Florida assembly when it drew flak from the Department of Natural Resources. Bob Ingle had urged the house members not to rush into such a law until the state could assess its economic impact upon the local fishermen who still landed green turtles and depended seasonally on this income. As a compromise, the legislature eliminated the three-year moratorium and replaced it with a mandate for the Department of Natural Resources, beginning January 1, 1971, to adopt minimum sizes for turtles taken or possessed in the state. When that version of the turtle bill was enacted, the job fell to Bob Ingle to come up with a minimum-size turtle regulation for the Florida green turtle.

Before Ingle's department could adopt the regulation, however, Governor Askew and his cabinet had to approve it. Under the Florida state constitution, the governor and his cabinet—composed of the elected secretaries of the departments of treasury, agriculture, education, and state, and the attorney general—sat as the policy-making body for all the departments of the executive branch. So Ingle and his boss, Randolph Hodges, had to present the proposed minimum size rule to a meeting of the cabinet's staff and then repeat the whole exercise before the full cabinet and governor.

Ingle was not looking forward to the exercise; it was likely to be just as contentious as the legislative committee hearings had been in 1970. In his view, Florida's environmentalists were waging a reign of ecological terror to bully the government into clamping down on all development and resource use.[29] Ingle believed his job was to keep the green turtle fisheries open as long as possible; his department depended heavily on support from the fish and game industries. But he had very little of the kind of biological information he usually had when he set fishing limits. The only information he had on the size at maturity of a green turtle was from his 1949 report with Walton Smith and anecdotal information from the fishermen. He tried to strike a balance between the biology of the sea turtle and the needs of the seasonal turtle fishery by setting the minimum size at twenty-six inches for west coast turtles. He knew that the men in the Cedar Key fishery rarely saw turtles larger than that. The green turtles that were landed in the only other fishery in the state, in the Florida Keys, seemed to average around thirty-six inches. Ingle decided to propose thirty-one inches as the minimum size for green turtles on the Atlantic coast, which would include the Florida Keys.[30]

When it was time for his department to present the proposed measures to the governor and cabinet, Ingle knew for sure there would be stiff opposition from the conservationists. Earlier in the month he had gotten letters from state senators asking what the scientific rationale was for his "26–31" proposal. He assumed that these letters were sent on behalf of Conservation 70's and Frank Lund, with whom he had tangled during the 1970 legislative hearings. Ingle was well aware that this young man had two things that he lacked: friends in the governor's office and data from his privately funded studies on Jupiter Island.

When Ingle got to the cabinet meeting, the room was filled with fishermen that the Southeastern Fisheries Association had bused in for the occasion to support Ingle's compromise proposal. The turtle item was moved to first on the agenda because five legislators were also there to voice support for the department's regulation. As soon as the item was introduced, Governor Askew began peppering Hodges with questions about the value of the fishery and the size of a sexually mature green turtle. Pressed repeatedly for specifics, Hodges was not prepared to present any data, hoping he could call on his director of research to field these questions. When the governor finally asked Ingle about the value of the fishery, Ingle said there were about seventy-five part-time fishermen in the Cedar Keys area who caught sea turtles seasonally, with a landed value of about $6,000. When they heard that dollar figure, several cabinet members turned and looked at each other with furrowed brows, wondering why it should take this much effort to deal with a regulation for such a miniscule industry.[31]

Still, the hearing dragged on as the cabinet members questioned the basis for the 26–31 proposal. Armed with tough questions prepared by the cabinet staff after its initial meeting on the proposed rule, the state treasurer, Thomas O'Malley, asked Ingle if he was the same Robert Ingle who wrote a report in 1949 citing a study from Australia that found an average size of thirty-five inches for a nesting green turtle. And if so, how would a thirty-one-inch measure protect green turtles in Florida? Ingle said the matter was complicated because turtles matured at different rates in the different regions of the world. Furthermore, he doubted whether Florida ever had had a major nesting population of green turtles; in 1949, he was unable to find an authentic account of a Florida population in the exhaustive literature search he had done for the report. Given this uncertainty, he was satisfied with taking an experimental approach. The 26–31 rule he proposed would achieve some conservation while allowing the fisheries to continue. Another cabinet member then asked Ingle how there would be any conservation if the law let the fishery catch every turtle before it was reproductively mature. Ingle said it was a bit of a gamble, but he was predicting that once caught and then released, the undersized turtles would not be caught again.

Governor Askew pressed him further for the data that supported his 26–31 proposal, asking "Doctor, could you tell me at what age a green turtle begins to reproduce?" Ingle replied that he could not, but offered to explain again his rationale for the proposal. State treasurer O'Malley asked if anyone had a scientific rationale, based specifically on the relationship between the length of the turtle's carapace and its reproductive maturity. When Ingle acknowledged that the length at reproductive maturity was probably above thirty-one inches, O'Malley asked him what was the wisdom, then, of setting a minimum size below that. Governor Askew asked why there should be a distinction between green turtles caught on the west and east coasts if the goal was conservation of one species with one age at maturity.

Ingle explained that his measures were an experiment to see if a size limit could slow down the depletion of green turtles in the state. The minimum sizes were designed to balance the biological needs of the species with the economic needs of the seventy-five people who fished for turtles in the summer months. A real solution to the possible extinction would need an international approach. He asked the cabinet and state legislature to ask the US Department of State to convene a conference of all nations bordering the Caribbean to discuss mutual conservation measures. He thought that the United States should join the tripartite turtle agreement, but that Florida fishermen should only stop fishing for green turtles when all the other countries agreed to do so: "Unless conservation is effected in the Caribbean, we're going to suffer for it in several species we've already identified."[32]

Governor Askew noted that the legislature was coming back into session in three weeks. His hope was to put an interim measure in place and then let the legislature do what it should have done in the first place and adopt a moratorium on all turtle fishing. He was frustrated by the lack of information the Department of Natural Resources had to support the measure. He asked if he could hear from Frank Lund, a university student whom he understood had some relevant information, so the meeting could move on to the rest of the agenda. Realizing it was now lunchtime, the governor then excused himself so he could attend an employee awards luncheon and gave the gavel to Secretary O'Malley.

Frank Lund then presented data he had collected since the age of nine on the sea turtles of Jupiter Beach, data that the Department of Natural Resources did not have. In twelve years he had never found a nesting green turtle that measured less than thirty-nine inches. That meant that the proposed minimum size of thirty-one inches would not protect a single green turtle from being killed before reproducing. In his view, the minimum size for turtles on the east coast of Florida should be at least forty-one inches. If a different measure was needed for the west coast to keep the fishery viable, he recommended twenty-nine inches, a length that corresponded to roughly 100 pounds, the size below which the fishermen believed a turtle had too little meat to be worth catching.

When Lund had finished his testimony, O'Malley asked if anyone else had any brief comments to make. Jack Rudloe from Panacea, Florida, came up to the microphone carrying a large box with a sea turtle in it. After reassuring the cabinet that it did not bite, Rudloe explained that he ran a business on the west coast of Florida collecting marine species for research laboratories and museums. He told the cabinet that green turtles were no longer present on the west coast of Florida. The last time he had caught a green turtle in Florida's Gulf of Mexico waters was 1964; the turtles that fishermen now were catching around the Cedar Keys and selling for soup were Atlantic ridleys, just like the one in the box. As he held up the female ridley turtle, he introduced her as "Little Bit" and explained how he had purchased her from a fishing boat three years ago to save her from the soup pot. She was the standard size that the fishery was catching. He bought as many turtles as he could from the fishing boats and sold them to aquariums to help promote Florida's wildlife. But he would gladly give up this part of his business to save the species. The fishery the department was trying to save was catching ridleys and loggerheads because the green turtles were gone. In his view, steps were needed to save the turtles, including more research and perhaps farming. He supported increasing the size limit to as large as possible until more could be learned about Florida's sea turtles.[33]

While Governor Askew was still absent, the cabinet members told Hodges that his staff should have taken this kind of testimony and resolved the question about the length of a mature turtle before coming before the cabinet. One remarked, "This is no place to establish the kind of facts that we've been trying to establish today." Before adjourning for lunch, the secretary of state moved that the regulation be changed to thirty-one and forty-one inches on the west and east coasts, respectively, based on Frank Lund's testimony and Ingle's agreement that a green turtle did not reproduce until it was at least thirty-two inches in length.

When the governor returned to the meeting, he gave the gavel to the secretary of state so he could make a motion himself and then vote on it. He proposed that a minimum size of forty-one inches on both coasts be adopted on an emergency basis, to take effect immediately. Once the motion was clarified, a roll-call vote was taken, and the forty-one-inch regulation was approved by a vote of five to one. The commissioner of education, Floyd Christian, cast the only no vote. Hodges and Ingle left the room stunned.

Shortly after the cabinet meeting, Ingle came very close to losing his job. The *Miami Herald* quoted him as saying the cabinet members voted for the forty-one-inch rule because they were afraid of the conservationists. Florida's attorney general promptly wrote to Randolph Hodges asking him whether the news story gave an accurate account of Ingle's views:

If it is an accurate report, I assure you my vote had no "political implications" and I doubt that other Cabinet members appreciate, anymore than I do, having a staff member ascribing our motives or intent, of which he could not possibly have any knowledge. Because, as the Governor stated, staff work was so bad that we could get no

FIGURE 26 Florida Governor Reuben Askew and cabinet meeting on proposed regulations.
Florida Photographic Collection, State Library and Archives of Florida.

information to support your recommendation, we had to rely on other professional guidance. I agreed with the Governor's suggested size limits and was glad to put his successful motion. I will appreciate hearing from you regarding post-action public comments by Dr. Ingle.[34]

In reply, Hodges told the attorney general that he had discussed the news story with Ingle. Hodges wrote that Ingle felt

that what he said to the reporter was misinterpreted. According to Mr. Ingle, he repeated to the reporter much of the testimony he gave to the Cabinet during the hearing on green turtles. This, of course, you are familiar with. In his behalf, I would like to tell you that he was better prepared for the hearing than his presentation indicated. I have directed him to prepare a detailed report on the present status

FIGURE 27 Marine specimen collector Jack Rudloe displays Kemp's ridley turtle purchased from fishermen outside his laboratory in Panacea, Florida. Photograph by Mike McClelland. Florida Photographic Collection, State Library and Archives of Florida, #PR20607.

of the sea turtle fishery. This will incorporate many of the facets of the problem that were not covered during the hearing. He hopes, as I do, that this report will be of value to all members of the Cabinet. We hope that his report will give a written documented presentation of his views, which will supersede any allegations that might have been made in the press or elsewhere.[35]

Three months later, Bob Ingle was, in a sense, vindicated when the Florida legislature passed a bill that established a twenty-six inch minimum size for turtles captured in the waters off Florida's Gulf coast, where the small commercial turtle fishery worked off Cedar Keys, and a forty-one inch minimum size for everywhere else, with an exemption for any green turtles that were caught in foreign waters and shipped into Key West. Governor Askew signed the bill into law on June 16, 1971.[36] But his signature came too late for the *A. M. Adams*. Its crew had gone home to Grand Cayman, and the turtle boat was now fishing for snappers.

CHAPTER TWELVE

One Man's Opinion

In March 1971, when Frank Lund was in Florida marshalling his best evidence for a minimum size rule while hoping for a moratorium, Archie Carr and the other marine turtle specialists were convening again at IUCN headquarters in Morges, Switzerland. Like the first meeting in 1969, this meeting was also funded by the World Wildlife Fund, which was closely associated with the IUCN. The specialists were now formally called the Marine Turtle Specialist Group, and Carr was its chair, with Tom Harrisson serving as co-chair. Harrisson was now living in Europe, divorced from Barbara, who remained in Ithaca, New York, to complete her PhD in anthropology. Harrisson divided his time between Brussels, where his new wife, Christine Fornari, lived, and the University of Sussex, where the Mass Observation archives were housed and where he was a visiting professor.[1] This arrangement allowed him to meet regularly with the IUCN Alert Group and promote the conservation of the many species he feared were disappearing, including the green turtle.

The Second Meeting at Morges

When each member of the specialist group gave a national or regional status report, Carr had the unpleasant task of reporting bad news from the Caribbean. The tripartite agreement for green turtles he had been so hopeful about at their meeting in 1969 had fallen apart. There would be no moratorium during which Costa Rica, Panama, and Nicaragua would develop a regional management plan. The Nicaraguan government had walked away from the agreement; the three turtle processing factories near Bluefields were in high gear, shipping thousands of turtle carcasses to northern soup manufacturers. In response, Costa Rica had allowed two factories to open near Limón.[2] Carr said the factories were part of the very troubling resurgence of trade in sea turtle products. The trade still focused on meat and calipee for soup manufacturing but had expanded to include turtle skins for exotic leather products, shells for so-called tortoiseshell, and oil for cosmetics. Even stuffed yearlings were being hawked to tourists at souvenir shops throughout the tropics. Late in 1970, the US Fish and Wildlife Service had listed the hawksbill, leatherback, and Atlantic ridley as "threatened with worldwide extinction," restricting their importation; but the green and Pacific ridley turtles could still be shipped to the United States.[3] With all seven species of sea turtles in the IUCN's Red Data Book, and three of them classified as endangered due largely to international trade, Carr hoped the draft global trade treaty that began with the 1963 Nairobi resolution would soon be concluded.

Antonio Montoya, the specialist from Mexico, reported that indeed the Pacific ridley was under enormous pressure from this trade. The official estimate of 750,000 killed in the 1970 season was an underestimation. The Mexican turtle leather factories were exploiting the ridleys' newly discovered group nestings, and ridleys were now disappearing. This was especially lamentable because these *arribadas*, as the ridleys' group nestings were called in Spanish, were an opportunity to study the role of social facilitation and breeding cues in migration.

Bob Bustard had somewhat better news from Australia. The Queensland government would soon expand its landmark turtle protection law to cover all species of sea turtles, including the population of loggerhead turtles, *Caretta caretta*, he had found nesting at a place called Mon Repos, near Bundaberg, Queensland. This beach also hosted a large nesting population of the flatback turtle, *Chelonia depressa*, which was an apparent cousin of the green turtles on the Great Barrier Reef. With such a restricted breeding range, the flatback needed serious taxonomic investigation as he had done for Harrisson's green turtles in Sarawak.[4]

Bustard was most excited about his news of the experimental green turtle farms on the islands of the Torres Strait, the tropical waters that run between the northern tip of Queensland and the southern coast of the Territory of New Guinea.[5] He explained that the native people of Queensland and the Torres Strait Islands had aboriginal turtle hunting rights and were exempt from the Queensland law. Unfortunately, commercial enterprises were abusing this exemption through uncontrolled hunting, to the detriment of the local people, who depended upon turtle and dugong meat. In the course of mapping all the sea turtle rookeries in Queensland, Bustard had taken the occasion to explain the situation to the Commonwealth Government of Australia and to propose a solution. The Commonwealth Office of Aboriginal Affairs gave him funds to initiate pilot turtle farms at a research level in the Torres Strait. The idea was to build up the green turtle population through restocking with captive-reared animals until the population reached the region's carrying capacity, at which point the local people would cull a carefully specified number through their traditional hunting methods. He had two goals for the project: to ensure the survival of green turtles at the extreme northern end of the Great Barrier Reef and to generate employment for the Torres Strait Islanders.[6]

When asked about the prospects for success, Bustard said they were enormous. He had spent quite a bit of time in the Torres Strait and the people there were eager to try farming, using eggs from the prolific rookery at Bramble Key. The people were delightful to work with; industrious and happy with their lot in life, all they required was a livelihood that would allow them to stay on the islands. Also, the Torres Strait Islanders were knowledgeable about the green turtle's ecology and life cycle. The men were superb and selective turtle hunters, and women knew when it was the best time to collect eggs from the nests. In fact, one young woman had developed a close, almost mystical friendship with an adult female she had named Ruth, and he hoped to write a book for the general reader about this friendship as well.[7]

Bustard was particularly pleased to report that the Commonwealth offered the young woman a six-month training grant at Australian National University in Canberra, where he worked. She would soon return to her island to begin raising green turtle hatchlings. Together, they would monitor the survival of captive-reared hatchlings released to the wild and get this all-important piece of information that had eluded both Carr's Operation Green Turtle and the Florida restocking program. They would closely record the culture practices that raised turtles most efficiently to a commercial size. He reassured the specialists that the modest number of eggs taken from the Bramble Key rookeries would be compensated for by the release of yearlings. Once they knew the survival rate of the released yearlings, they could adjust the number of wild eggs taken for the farms to ensure that the appropriate ratio was taken.

Bustard also described his experimental crocodile farm in northern Queensland, an outgrowth of his long-standing interest in crocodilians. He had recently founded the IUCN's Crocodile Specialist Group, on which he had the pleasure to work with Wayne King and Duke Campbell, whom he knew to be close associates of Archie Carr's. But he hoped that Carr would be especially interested in the farms on the Torres Strait Islands. They were the cottage-scale projects that Carr had suggested as early as 1963, in his book *The Reptiles*, and had elaborated on in *So Excellent a Fishe*. Bustard noted as an aside that he also had a book in press for the general reader on the natural history of sea turtles. In it he too made the case for farming as a means of conserving those species of sea turtles that were heavily hunted for food and other products.[8]

When the technical and conservation aspects of turtle culture came up on the agenda, John Hendrickson gave a paper on the research he had begun while still in Hawaii. Before joining the biology department at the University of Arizona, Hendrickson had been for two years the director of the Oceanic Institute, a mariculture research and development facility on the island of Oahu, which owned a captive display facility called Sea Life Park. There, he had begun studies on the growth and nutritional requirements of captive-reared green turtles from the Hawaii population. He obtained hatchlings from the US Fish and Wildlife Service, which oversaw the Hawaiian Islands National Wildlife Refuge, where green turtles still nested in reasonably large numbers on the islets of French Frigate Shoals. Now that he was in Arizona, he hoped to continue this type of work on turtles from Mexico, perhaps from the Gulf of California.

Hendrickson told the group that the nutritional studies might be continued at Sea Life Park. As he had mentioned at the meeting in 1969, he had urged the Sea Life Park investigators to set up a small-scale farm in a bay or lagoon where they could work out standard techniques for rearing young green turtles to market size. He was confident that green turtles would lay eggs in captivity—a necessary condition if farming was to aid conservation. He had gotten the green turtles in his care to lay eggs in a pool with an artificial beach.[9] To cover some of the expenses, he encouraged the researchers at the park to enlist the aid of the new luxury tourist hotels springing up along Waikiki Beach. These hotels served turtle steaks and soups and, he thought, should be very receptive to the idea of raising their own turtles.[10] In the meantime, Hendrickson was about to embark on the turtle culture feasibility assessment for the South Pacific that the IUCN's Harold Coolidge had encouraged him to do. Hendrickson would soon visit several islands, including Fiji, which was of particular interest because Bustard had found the turtle population in such dismal shape on his last visit.[11]

The discussion of turtle farming took on a more critical tone when several guests of the meeting were introduced. They were directors of Mariculture, Ltd., the commercial turtle farm on Grand Cayman Island, who had invested in Robert Schroeder's project and were now seeking the approval of the sea turtle specialists. Joining them was a representative of Lusty's Soups, one of Mariculture's major customers. The discussion focused on whether turtle farming would make a positive contribution to conservation of wild populations. The directors made the case that demand for sea turtle products was already high, too high to be sustained by wild populations. Their plan was to satisfy this demand with farm-raised turtles and in essence put the poachers out of business. Their product would be far superior to wild-caught turtles, and the companies they sold to would be able to advertise their product as an environmentally sound alternative to wild turtles.

Carr asked pointedly if they really believed they could undercut the sales of the Nicaraguan factories. They replied that of course they did. They were particularly sanguine because the IUCN's draft treaty on trade in endangered species was nearing completion. Once it was adopted and in effect, any country that signed the treaty would be unlikely to import wild-caught green turtle products from Nicaragua or anywhere else. The treaty would not apply to Mariculture's exports because these products would be farm-raised. Tom Harrisson noted that Barbara Harrisson had been following the draft treaty closely on behalf of the primate specialists and it looked like all the species in the Red Data Book, including the marine turtles, would be on the trade treaty's list.

Some of the specialists wanted to know how Mariculture could claim it produced an environmentally superior product if the company was taking eggs from the principal wild green turtle rookeries at Tortuguero, Ascension Island, and Suriname. The directors responded that under an agreement with the Costa Rican agriculture ministry, they were returning captive-reared one-year-old turtles and releasing them on the rookery beaches. They understood from the scientific literature that because of natural predation, hatchlings stood only a 1 percent chance of surviving to adulthood, while yearling turtles had a much higher likelihood of surviving. By returning 10 percent of the turtles that had been taken as eggs from the beach in the form of yearlings, Mariculture was more than compensating the population for the farm's egg collection.

Joop Schulz mentioned that there would soon be similar programs in Suriname and Ascension. Schulz, the Dutch zoologist with the Suriname Forest Service, had started his own nature conservation organization, and this program would soon start selling green turtle eggs to Mariculture in exchange for monies to pay for operating the turtle hatcheries. The Forest Service was not planning to increase the number of eggs it already allowed to be taken for human consumption; the Mariculture allotment would come from this number. All the eggs Suriname sold to Mariculture would be those that would not hatch if left in situ owing to the extensive erosion at Bigisanti Beach.

The man from Lusty's Soups said he was so impressed with this conservation program that his company was going to sign an exclusive supply contract with Mariculture and use its turtle calipee to make turtle soup for the lord mayor of London's banquet.[12] After much discussion about the assumptions underlying their conservation program, the Mariculture directors agreed that until they were producing their own eggs on the farm, they would obtain only those eggs that would not otherwise hatch due to conditions on the beach. The turtles hatched from these "doomed eggs" would be used for their breeding stock rather than for production; Lusty's and Mariculture's other customers would get only products from farm-raised turtles, not from the wild. Lusty's representative agreed that he would continue to refuse "to buy turtles that had been illegally or inadvisably taken."[13] He would also put the words "farmed not wild turtle" on the label so his customers would have confidence they were not endangering wild turtles. Although several of the scientists attending were still unsure about the feasibility of turtle farming and a bit skeptical of these promises, the specialists agreed to share information from any research they did on captive culture techniques. Mariculture's directors extended an invitation to everyone in the specialist group to come and do research at their facility.

The discussion of farming's potential contribution began to reveal a serious division of opinion among the specialists, which seemed exacerbated by the lingering resentment by one or two of them that Carr's close associate, Peter Pritchard, had been selected as the marine turtle group's executive officer. In private conversations, these disaffected specialists convinced Leo Brongersma and John Hendrickson to ask the IUCN officials if Pritchard was the best person for the position and suggest they replace him with someone who was broad-minded on the issue of commercial exploitation. Colin Holloway, executive officer of the IUCN's Survival Service Commission, raised the issue at the end of the meeting under the agenda item regarding future directions. Tom Harrisson was appalled by the apparent "careerism" underlying this attempted coup, but Carr was just sad and very embarrassed. When Bustard, one of the specialists who had raised questions about the executive officer position at the 1969 meeting of the group, saw the discussion was not going well, he tried to cover his tracks by turning on Hendrickson and Brongersma, asking how they could even think of such a thing, offering to flagellate himself and others with his rhinoceros hide whip.[14]

Before the challenge to Pritchard's appointment ended the meeting on a sour note, the specialists were able to reach agreement on research priorities for the next two years. These were (1) mapping of group nesting sites, particularly in areas previously undocumented in Australia, Asia, Indonesia, and South America; (2) developing captive culture techniques; (3) conducting population ecology and energy-flow studies; and (4) conducting taxonomic-zoogeographic studies. The members agreed that it was appropriate for the group to serve as a special consultative body to governments in those places where turtles occurred and to offer their counsel on every aspect of turtle conservation. They also agreed to condemn the growing sales of stuffed yearling turtles as souvenirs but to note that the uncontrolled collection of turtle eggs was the greatest threat because this practice affected all species. They condemned the leather trade for its impact on the ridleys, and the vandals who were senselessly killing leatherback turtles.[15]

After the meeting, Leo Brongersma invited Bustard and Pritchard to lunch, and Pritchard made it clear he had no hard feelings about Brongersma's unwitting role in the embarrassing matter of his appointment.

Pritchard thought Brongersma was such an excellent conversationalist, it was impossible to feel anything but delight in his company. Bustard, however, did not say a word, and Pritchard wondered what he was so angry about. They had disagreed over Pritchard's assessment of the conservation status of the leatherback, but that was no reason to be so dour.[16]

The Last Straw

At the specialists' meeting, the directors of Mariculture had invited John Hendrickson to come to the farm and help Bob Schroeder with the culture methods. When Hendrickson got back to the University of Arizona, he asked his graduate student Jim Wood if he would like to work on the amino acid requirements of green turtle hatchlings for his dissertation instead of red snappers. Wood readily agreed.[17]

Upon returning to Gainesville, Carr heard all about the flap over Florida's forty-one-inch rule and the end of the "turtle boat." Wright Langley, from the *Miami Herald*, called and asked Carr what he thought about the new minimum size regulation. Carr told him he felt sympathy for the turtle fishermen because some of them had helped his studies. Would the new regulation mean the green turtle would be less endangered? Carr told him that forty-one inches was not a magic number and that all green turtles really should be given a breather, especially the larger ones. He said, "The old females who laid most of the eggs are the ones that should be spared to keep the population going." Trade in all sea turtle species was on the rise worldwide and he feared every population would soon be in the same shape as the Florida green turtles.[18]

In reviewing his notes from the Morges meeting, Carr was annoyed to recall the boastful promises by the men from the commercial turtle farm on Grand Cayman. Mariculture, Ltd., had grandiose plans to franchise its turtle culture techniques and spread turtle farms throughout Latin America and perhaps the world. But despite Schroeder's best efforts, the company had not yet managed to breed turtles in captivity, much less raise any to maturity. To Carr, the vision of commercial green turtle farms throughout the tropics seemed as big a threat as the slaughterhouses and leather factories in Nicaragua and Costa Rica. It was particularly galling to be reminded at every juncture that it was his discussion of farming in his popular writings that had inspired Bob Schroeder and the investors in Mariculture. The cottage-scale turtle farms Bustard was developing on the Torres Strait islands were closer to what he had had in mind, although Bustard's idea of marketing the turtle products was troubling.

Then Carr received a letter from IUCN headquarters asking his views on a request by Mariculture for World Wildlife funds to support their research. It was the proverbial last straw. He replied: "For a wealthy company such as they are to take foundation money that would otherwise support the work of conservation and research people seems unreasonable. As to the 'banks' of information and of genes that they propose to establish, we all are trying to establish the former; and as for the latter, I believe we are not quite at the stage that we have to save turtle species in cages." He explained that Mariculture had been told repeatedly that before conservationists would regard it with favor, the company would have to make itself completely self-sufficient and no longer take eggs out of nesting sanctuaries "under any kind of formula or arrangement"; moreover, it should start production only when it could show that it could undersell the market. Neither of these conditions had been met. The company had decided to go straight into production, and now that it had run into technical problems, it was asking for research money from charitable foundations to solve them. "I honestly don't understand the reasoning involved at all," Carr concluded.[19]

Carr decided it was time to put his current view of commercialization, especially farming, down on paper. *Audubon Magazine* had invited him to become a contributing editor and to write a hard-hitting article on why each species of sea turtles was endangered. This would be an opportunity to set the record straight on farming and at the same time to speak out against the growing vogue of turtle steaks and fashion. On its own, the article would not stop commercialization or prevent the farms from proliferating, but at least everyone would know where he stood.

Great Reptiles, Great Enigmas

In his article in *Audubon*, published in March 1972, Carr made it clear that the oceans, long thought unassailable, were now in decline. But no group of species in the sea was "suffering from human intemperance" more than the sea turtles. Three species were listed in the IUCN's Red

Data Book as endangered, "and by the criteria the IUCN employs that is a reasonable judgment." But if trends are used as the measure of peril rather than fixed population levels, he wrote, "there is no cause for complacency over the survival outlook of any kind of marine turtle."[20]

He then catalogued the threats to each species and attempted to rank them in order of endangerment, much as he had done for the Red Data Book. Writing for *Audubon*'s popular audience gave Carr greater scope for editorializing on the nature of the threats and the efficacy of the current countermeasures. He noted that in his 1952 *Handbook of Turtles* he had predicted the extinction of the green turtle if there was any resurgence of the turtle meat industry that had driven the "once teeming hordes" into a "pathetic remnant." That resurgence had now occurred, and the governments responsible for protecting the green turtle were failing to do so.

He described the collapse of the international agreement for a three-year moratorium and the scale of the modern processing plants in Nicaragua and Costa Rica. These processors were exploiting the Tortuguero population into oblivion, despite knowing that the turtles caught in Nicaragua were from the last remaining green turtle breeding ground in the western Caribbean: "This simple ecological picture has repeatedly been made plain. And still they continue to subject the declining resource to growing exploitation."

In making the case for placing *Chelonia mydas* in the antepenultimate place in his own list (behind only the Atlantic ridley, *Lepidochelys kempii*, and the hawksbill, *Eretmochelys imbricata*, which were tied for most endangered), he noted that even in Florida, "medieval legislation, staunchly supported in recent political controversy, permits the commercial exploitation of an almost vanished green turtle colony." He was referring to the fact that three months after Florida's governor and cabinet adopted Frank Lund's forty-one-inch minimum size, the state legislature had enacted a superseding law allowing the west coast turtle fishery to take turtles of twenty-six inches or greater and to exempt green turtles from the minimum size if certified by the dealer as foreign-caught.[21]

Carr reviewed the knowledge that he and his team, including his wife, Marjorie, had pieced together in the almost twenty years of tagging and measuring the nesting population at Tortuguero, and explained how this knowledge was slowing revealing the mysteries of the turtles' growth and migration cycles. Thanks to Marjorie's statistical analysis, they now knew that any turtle with a carapace length of thirty-six inches that turned up at Tortuguero was more than likely to be a new addition to the nesting colony, and that nesting cycles varied. Some took two years and others took three to return. He discussed what he now believed about site fixity and navigation and the frustrations of his attempts at electronic open-sea tracking.

By summarizing his work Carr was reminding his readers in *Audubon* why turtles should be saved: what science could learn about their adaptations was of far more value than a turtle steak dinner in a tourist-trap restaurant. He described the tactical actions that might buy sea turtles more time, while waiting for international cooperation at the governmental level—"an almost legendary substance"—to materialize. These steps were "more sanctuaries, more research, and a concerted effort by all impractical, visionary, starry-eyed, and anti-progressive organizations, all little old ladies in tennis shoes, and all persons able to see beyond the ends of their noses, to control the international commerce in sea turtle products."

Without identifying it as such, Carr described the specialists' recent debate at Morges on "whether to mount international campaigns to discourage the use of products derived from the animals, or to encourage artificial culture products." But these were not incompatible actions in his view, notwithstanding the complaints of prospective green turtle farmers. The farming entrepreneurs argued that if conservationists really wanted to save green turtles, they should cease their attempts to kill commerce in turtle parts and instead embrace turtle farming. Once trade in domesticated turtle products was established, entrepreneurs argued, its efficiency and superior product would put the poachers and commercial turtle hunters out of business.

Carr made it clear he was not buying this argument. "Superficially it sounds good, but there is fuzzy thinking in it. I have yet to see or hear of a work plan for any reptile ranch that shows in realistic detail how it expects to achieve a volume of production so great that it will do anything other than *increase* both demand and prices." A commercial turtle farm was bound to create new markets but fail to satisfy them, only exacerbating the situation of wild populations. He was aware that individual female green turtles had mated and laid eggs in captivity, but production-scale egg-laying was many years away. In order to "flood the market" with farmed turtles, farmers would have to raise little turtles to slaughter size from eggs collected on the breeding sanctuaries, further draining the wild population.

People who were truly interested in the survival of natural species should advocate only one form of turtle farming as an alternative to depopularizing turtle products. That would be a program run by a non-profit, government-sponsored entity in which "many small, widespread, purely experimental projects simultaneously attacked the problems of nutrition, disease control, and captive breeding," and which from the beginning freely shared information on its procedures and results. In describing everything that an acceptable pilot farming program should be, Carr had described everything that he believed Mariculture, Ltd., was not.

Sea Turtles on the Legal Ark

As Mariculture's directors had told the turtle specialists in Morges, the company's business plan depended on an international law that did not yet exist. A treaty banning commerce in wild-caught sea turtles would give their farm-raised product exclusive access to the lucrative global markets for turtle shell, meat, oil, and skins. From his estate in England, Antony Fisher kept in touch with IUCN officials on the treaty's progress. He found the pace excruciatingly slow for an idea that had received unanimous endorsement at the IUCN's Assembly in Nairobi.

Since 1963, the IUCN had drawn up and reviewed no fewer than four draft versions of the treaty.[22] In each draft, the core idea remained the same: to ban or control trade in wild animals and plants that were on global lists based on the Red Data Book.[23] As reviews of the drafts trickled in, the IUCN realized that "endangered species" lacked a common, consensual definition. Exporting and importing countries differed on who should decide which species belonged on the list.[24]

In the 1969 Endangered Species Conservation Act, the US Congress directed the secretary of state to seek an international agreement on wildlife trade so that the US fur and skin importers and manufacturers would have a "level playing field" with their competitors in other countries. But the two-year deadline for that treaty had passed, and countries were still arguing whether there should even be a global list, based on the Red Data Book or some other authority, or whether the treaty should simply rely on national lists. Another target date was the United Nations Conference on the Human Environment in Stockholm in June 1972, but the draft treaty was not ready for adoption then either.

Finally, officials in the White House Council on Environmental Quality and in the US Department of the Interior revised the draft treaty to combine the global and national list approaches and invited 100 countries to an international conference at the State Department from February 12 until March 2, 1972. The main job of the delegates to the Plenipotentiary Conference to Conclude an International Convention on Trade in Certain Species of Wildlife was to put the final touches on the document and decide which of an expanded list of animals and plants were threatened with extinction due to international trade. This was where the haggling was sure to enter in.[25]

The delegates' job was to approve the placement of animal and plant species on one of two draft lists. The first list included those species considered by IUCN to be threatened with extinction which currently were or could be affected by trade. Species on the second list were deemed not currently threatened with extinction but could become so unless trade was strictly regulated. Exporting countries could authorize trade in individual specimens of species on the first list only for scientific or conservation purposes and then only if the importing country also certified that the trade would not threaten the species' survival. For species on the second list, the country of origin could issue an export permit if it deemed that trade in a specimen was not detrimental to the survival of the species.[26]

The Interior Department appointed Wayne King, the curator for reptiles and director of conservation of the New York Zoological Society, to the US delegation. King was still very active in endangered species policy circles, and he chaired the conference's reptile working group. His main goals were to make sure that seven species of crocodilians remained on the first list and to transfer the American alligator from the second list to the first. Also present at the conference, though not as a delegate, was Frank Lund. Lund had moved from Florida to Washington for an internship at the Interior Department, where fellow Jupiter Island resident and family friend Nathaniel Reed served as assistant secretary for fish, wildlife, and parks. Lund sat in on the reptile working group negotiations to see what would happen to the three marine turtle species on the first list and the three on the second.[27]

Lund was very impressed by the skill with which King handled the negotiations. The most heated discussions concerned how endangered the American alligator was and whether international trade or habitat loss was the major factor. Compared with these arguments, the placement of the sea turtles seemed a cinch.[28] The most hotly debated question was where to place the hawksbill turtle. Japan opposed including it on the first list because of Japan's extensive trade in hawksbill shells for the manufacture of tortoiseshell luxury goods such as jewelry and art objects.[29] For entirely different reasons, the Australian delegate did not want the green turtle on either list. Australia thought the 1968 Queensland legislation was a model other nations should emulate. Listing would imply that Australia had not done enough to control exploitation, although, in fact, the government had promoted exports to England and Europe in a previous era. This trade had certainly depleted green populations on the Great Barrier Reef.

By and large, matters were moving along on the reptile working group, but the committees negotiating the mechanisms for the trade controls were at odds because of differences between the exporting and importing countries. Halfway through the plenipotentiaries' conference, these differences appeared irreconcilable, and a worldwide treaty to protect endangered wildlife seemed just out of reach. There were endless debates over points of semantics, with nations offering amendments to clarify the text or to add further ambiguity to support their positions.[30]

The tenor of the discussions changed markedly when Lund's friend Nathaniel Reed, the assistant secretary of Interior and a member of the US delegation, made a brilliant move. He held a press conference to announce that US Customs and Fish and Wildlife agents in New York had just broken up an international fur-smuggling ring that involved the New York furrier Vesely-Forte, Inc., and thirty-two co-defendants. The ring operated on four continents and accounted for roughly half of the entire US market in spotted cat skins, including leopard, ocelot, cheetah, jaguar, and puma. In less than two years, this group of poachers, smugglers, and dealers had managed to sell the skins of 10 percent of the world's remaining cheetahs. This revelation made clear what would happen to the world's wildlife if a treaty could not be agreed on.[31]

By the end of the conference, the only sea turtles on the first list, which was now called Appendix I, were the Atlantic hawksbill subspecies, *Eretmochelys imbricata imbricata*, and the Kemp's (formerly Atlantic) ridley species, *Lepidochelys kempii*. All other sea turtle species and subspecies were on the second list, now called Appendix II, including the Atlantic and Pacific green turtles.[32]

Nathaniel Reed and fellow US delegate Lee Talbot were happy with the outcome but realized that their next battle would be to get Congress to amend the endangered species legislation so that they could implement the treaty in the United States, one of the largest markets for wildlife and its products. Given the difficulties the Interior Department had in listing species like the great whales that were still commercially exploited in the United States, Reed and Talbot drafted a new version of the legislation. It was titled the Endangered Species Act of 1973, and President Nixon had introduced it in his State of the Environment address in February. Like the new treaty, the new ESA would recognize two categories of endangerment: threatened with extinction and likely to become threatened with extinction. Under the new law these categories would be called "endangered" and "threatened." It would no longer be necessary for the interior secretary to determine that a species was "threatened with worldwide extinction" in order to put it on the US list. Also, the secretary could list and protect a subspecies or geographically isolated population if it was in trouble. The decisions on where to place species would be made based on the "best available scientific information."

After the conference ended and the U.S. had signed the Convention on International Trade in Endangered Species of Wild Flora and Fauna, or CITES, as the treaty was called, the staff of the endangered species office at the Interior Department realized they should list all the Appendix I and II species under the ESA. Only then could the United States meet its obligations to control trade in the treaty-listed species. From discussions at the conference, it was clear that particular species needed to be listed as soon as possible and their imports to the United States banned. Among these species were the sea turtles. Although there was much to do to prepare assistant secretary Reed for congressional hearings on CITES and the bills amending the endangered species law, Interior staffers quietly published the most urgent and long-overdue listings.

The process took much longer than they had planned. Few people noticed when, on December 28, 1973, the Interior Department finally published a proposal to list the green turtle on the US list of endangered foreign fish and wildlife. The accompanying press release said that green turtles, "once believed to have numbered at least fifty million, now are estimated at less than ten thousand. Reproductive potential may be destroyed in the near future if present harvest levels are maintained."[33] Interested persons were invited to submit comments or objections.[34] As luck would have it, the proposal was published the same day that President Nixon signed the new Endangered Species Act of 1973 into law.

Wayne King saw the proposal, however, and discussed it with Frank Lund and Archie Carr early in the new year. Carr asked whether once the listing was final, the turtle farm on Grand Cayman would be unable to send its ghastly turtle products through the port of Miami. Would it be a valid listing under the newly signed legislation? The answers were not apparent. Even if the green turtle remained on the US "endangered" list, trade in endangered species was still possible under the new treaty, King explained. In a compromise between delegations from exporting and importing countries, the parties had added some potential loopholes to the CITES text in the negotiations. The strict permit system would not apply to transshipments, specimens or products that were acquired prior to the treaty's application, or to products made from animals "bred in captivity for commercial purposes."[35] Would that apply to the farmed green turtle products? He did not have an answer.

CHAPTER THIRTEEN

Down on the Farm

A s soon as he could get a copy of the Interior Department's proposal to list the green turtle as endangered, Archie Carr sent a letter of support to the Fish and Wildlife Service. Pressed for time, as he always was, Carr kept the letter short. He thought it was probably unnecessary to include data to support the endangered classification. At any rate, he had time only to summarize the condition he believed the green and loggerhead turtles were in.

> I consider this to be a wise move, already seriously overdue, and badly needed to strengthen efforts to stem the accelerating decline of the green turtle and Atlantic loggerhead. *Chelonia* is suffering everywhere from the expanding markets and rise in prices that have been stimulated by culture projects and by the increasing prosperity of some consumer nations. *Caretta* is being dragged down by the loss or disruption of its habitat, by virtually uncontrollable man-induced raccoon predation, and by the increased accidental drownings that occur in the enormous shrimp trawls now in use. While neither of these species has reached the low absolute population numbers of a whooping crane or a blue whale, if ecologic and exploitational trends are recognized as evidence, then both are clearly in terminal decline. I therefore hope you will do everything possible to implement the [proposed] amendment to give them Endangered Status.[1]

As he had done with the Red Data Book pages in 1967, Carr made it clear that it was these trends that made extinction a real possibility. For the green turtle, farming was producing demand for luxury items for the new consumer society.

Call in the Biologists

Carr's view of the conservation benefits of turtle farming on any scale was growing dimmer by the hour. He was annoyed by all the lobbying that Mariculture, Ltd., had subjected him to throughout the summer and early fall of 1973 as the company tried to get California and then New York to exempt its "farm-raised" sea turtle products from their faunal protection laws.

Carr had learned from John Hendrickson that Mariculture now had a department of conservation and research chaired by a very distinguished scientist from England, but he thought it was a little late to be just now figuring out how to make a captive turtle herd breed enough to keep up with the skyrocketing demand. In a brochure released during the summer of 1973, entitled "New Hope for the Green Sea Turtle," Mariculture had boasted it had already increased the population of green turtles in the Caribbean by releasing "head-started" yearling turtles raised on the farm from wild-laid eggs collected from Costa Rica, Ascension Island, and Suriname.

In June 1973, Hendrickson had called Carr to tell him that good scientific research was being done at the farm to work out the thorny problems of reproduction and nutrition. Hendrickson now had two graduate students, Jim Wood and Dave Owens, working hard on these problems, alongside two eminent biologists whom Mariculture had recruited from England. One of the biologists was Emmanuel C. Amoroso, whom Antony Fisher knew from his Buxted chicken company. Amoroso was an emeritus professor of reproductive physiology from the Royal Veterinary College of the University of London and a Fellow of the Royal Society. Amoroso in turn had recruited Sir Alan Parkes, another Royal Society Fellow, recently retired from the University of Cambridge and an expert in reproductive biology. On one of his vacations to Grand Cayman Island, Parkes had seen the farm in the early stages of its expansion and was now serving as chair of its science department. "Amo" was working on problems of incubation, embryonic development, and mortality, while Sir Alan tackled the breeding and husbandry challenges.[2] Parkes had the idea of introducing two wild adult males from Suriname into the breeding pond. When they did so, all the turtles in the unsuccessful breeding pond went into a frenzy of mating. By the end of

FIGURE 28 Mariculture, Ltd., breeding pond and artificial nesting beach at Goat Rock, Grand Cayman Island.
Courtesy of David William Owens.

May 1973, almost 11,000 eggs had been laid. The farm had succeeded in the first-ever captive breeding of sea turtles.

Carr asked Hendrickson how many of the eggs produced viable hatchlings, as that was the key to ending the farm's dependence on eggs taken from natural rookeries. Hendrickson told him that the hatch rate was close to 43 percent but was twice that for nests that had stayed in the farm beach and had not been moved to the hatchery. Carr said he thought it was a fine thing that Hendrickson and his students were tackling the biological questions. His main objection was that at the same time this great work was being done, the marketing people were at full tilt, expanding markets for farm-raised products that were still only being produced from wild turtle eggs.[3]

Carr was chagrined to learn from Hendrickson that Mariculture had made a deal with the Bergen cosmetics company in California. He realized that Polly Bergen, like Bob Schroeder a decade earlier, had misinterpreted his musings that perhaps green turtles should be cultured. Already

FIGURE 29 Meeting of Mariculture staff and consultants on captive reproduction. From left to right: Glenn Ulrich, research manager; Julie Booth, consulting behavioral ecologist; Mike Goodier, managing director; Sir Alan Parkes, reproductive biology consultant (standing); Professor E. C. Amoroso, reproductive biology consultant (seated); and Irvin Naylor, president of Mariculture, Ltd. (standing).
Photograph by John R. Hendrickson, consulting turtle biologist for Mariculture; used courtesy of David William Owens.

a successful actress and recording artist, Bergen started a cosmetics company in 1965 with a face cream product called Oil of the Turtle using by-products of the manufacture of turtle soup; the company also sold tortoiseshell jewelry. A turtle fancier, Bergen joined the Caribbean Conservation Corporation as a lifetime member after reading Carr's *So Excellent a Fishe.*

In 1970, Bergen had written to Carr, asking if traffic in turtle oil was illegal and hastening the green turtle's extinction. Carr answered bluntly. The populations were already overharvested, and the new demand for oil and leather was "raising the price on the heads of the turtles," encouraging both legitimate agents and illegal poachers, and vastly complicating the enforcement problem. He hoped the new turtle commerce would fail for lack of resources before the green turtle went extinct. Carr told Bergen that if he were in her shoes, he would either get out of the business or "rigidly curtail" her output and make "desperate efforts to develop techniques of turtle culture—religiously keeping these at a small-scale experimental level until all major procedural problems had been solved, and only then undertaking to begin commercial production and expand markets. . . . In a civilization as technologically advanced as ours, dependence on wild animals for any commercial product is anachronistic. Certainly no fad should be allowed to threaten the existence of irreplaceable forms of life."[4]

Bergen apparently believed that Mariculture had resolved all the technical problems, and Carr suspected that the company's advertisements had led her to believe this. When Antony Fisher wrote Carr in July 1973 and asked for the support of the Marine Turtle Specialist Group in proceedings before the California assembly, Carr decided it was time to tell them what he really thought and why. Fisher should not expect endorsement from the species-survival community, Carr wrote, until it produced a "realistic prospectus showing (1) projected evolution to complete self-sufficiency with respect to the egg needs within a short time; and (2) a predictable production schedule that will saturate markets for turtle products—meat as well as the various by-products of your enterprise—and depress prices to the point that both poaching and legal turtling are made unprofitable. Failure to meet these stipulations automatically makes Mariculture one of various adverse factors in the survival outlook of the green turtle."[5]

Carr's patience was further tried when later that summer Hendrickson wrote to ask his views on a legislative proposal Hendrickson had made to the California assembly that would allow Mariculture to stay in business. The company's exclusive supply contract with the Bergen cosmetics company was in serious jeopardy unless a bill pending in the California assembly passed, allowing Mariculture to get a permit to sell turtle products in the state. This bill would, Carr knew, create an exemption from the faunal protection law that Wayne King had worked hard to get enacted in New York and California to stop the fashion for alligator shoes and ladies' fur coats from driving the alligator and big cats to extinction. Mariculture was lobbying for an exemption to sell green turtles for an antiwrinkle cream.

Hendrickson told Carr he had testified in favor of the bill for two simple reasons. First, the farm was close to making a breakthrough in captive reproduction, a feat that would open up whole new avenues of research that would contribute to sea turtle conservation. It needed to sell oil to companies in California to stay afloat while the research department worked out the final kinks in the reproductive cycle. Second, he had a personal conviction that "appropriately organized and regulated sea turtle farming activities can, in the balance, help the cause of sea turtle conservation more than they hinder it." Hendrickson had suggested to the legislators that their law authorize a five-year renewable permit allowing sales in California with two conditions: first, that until the farm was producing sufficient quantities of its own eggs, Mariculture would use only wild eggs that were doomed—that is, would not hatch in the natural nest because of its location—verified through documentation and inspections; and second, the farm would be required to take "positive conservation action" such as funding projects like the hatchery and anti-poaching patrols at Suriname's nesting beaches. The farm would need only a one-year grace period to meet these stipulations.[6]

Hendrickson asked Carr for his thoughts on these proposals and suggested that they visit the farm together, when they could both talk to Sir Alan. But Carr had neither the time nor the inclination to answer. He had said all he planned to say on the subject in the letter to Antony Fisher that Hendrickson would probably see soon enough. But when a consultant working for the California Assembly's Committee on Natural Resources wrote to Carr to ask for his endorsement of Hendrickson's proposals, the man made the mistake of suggesting that Carr's unwillingness to visit the farm to assess firsthand its contribution to conservation was unfair.

Carr sent him a terse reply, telling him to read the letter Carr had sent to Antony Fisher, and admonishing him for his implication, writing: "As for your astonishing suggestion that my failure to visit Mariculture is somehow reprehensible, I think your enthusiasm for that enterprise is getting a little out of bounds there. I am not an opponent of Mariculture's. I do, however, reserve the right to reserve my praise of the organization until I have seen that it is not impairing the status of the wild species that for 20 years have been my special concern. As of now that is simply not clear."[7]

As to businessmen's mentality, he did not have to talk with them to understand that but could rely on David Ehrenfeld's assessment. Ehrenfeld had met some of the principals of Mariculture, Ltd., the previous winter when they attended a talk on mariculture that he had given at the New York Academy of Sciences.[8] Carr agreed with Ehrenfeld that large-scale commercial turtle farming made no sense ecologically, but the Mariculture directors did not agree. As investors and entrepreneurs, their job was to make money, not to worry about the big picture.

Dire Straits

Carr, in fact, was getting ready for a trip to see a different kind of turtle farm. He was heading off to the islands in the Torres Strait. Bob Bustard's experimental turtle farms had run into political troubles, and the Australian government had asked Carr to investigate the ecological implications of his farming scheme as part of an inquiry into the farming project. Carr was eager to meet the Torres Islands turtle farmers. These were turtle people like the Miskito villagers of Tortuguero and Tasbapauni. He had not yet been to that part of the world, and there were wild sea turtles there in abundance.

As Bustard had reported to the meeting of the marine turtle specialists, he had formed the Applied Ecology Unit at the Australian National University and set up a network of cottage-scale turtle farms, raising hawksbills and green turtles for local consumption and for trade. Although Carr was tiring of the farming controversy, he had never been to the Torres Strait, and he wanted to see the Australian flatback turtle currently called *Chelonia depressa*. The taxonomy of this purported cousin of *Chelonia mydas* needed to be worked out, and Bustard might make a hash of it if Carr did not help.

Unlike the requests Carr received from John Hendrickson and the consultant to the California legislators, those he received from the Australian politicians had asked him to assess the ecological implications of turtle farming. This he felt comfortable doing. He would not have to say whether turtle culture would lower the price enough to flood the market and discourage poaching or would drive up demand and increase poaching. That question was beyond his expertise, yet everyone expected him to venture an opinion. No tour of the farm on Grand Cayman, nor meeting with their hired guns, could inform his views on that question. He could only guess from what he knew of human appetites, and he certainly knew how efficient the calipee poachers were from his years of encountering their handiwork on the beach at Tortuguero.

Carr enjoyed the trip, especially the company of the captain and crew of the *M. V. Moreton* that carried him and his inspection team to nine of the tiny islands in this remote archipelago, all but two of the islands where turtle farming was under way. Carr and Albert Main, a professor of zoology from the University of Western Australia, interviewed over 100 farmers and inspected their turtle rearing facilities. Although the business organization was beyond Carr and Main's remit, Carr could see problems with the plan.[9] Each family owned its own farm, and each farmer was given a training allowance only after he or she had at least 150 hatchling turtles. They were required to increase the number to at least 250 in the first three months. When the turtles were large enough, the plan was to put them in two or three large grow-out pens to be built in the shallow water. These pens were to be commonly owned and operated by an island-based company that purchased the young turtles from the individual farmers.

On the way to Australia, Carr had read the preliminary draft of an ecological impact statement prepared by Bustard's Applied Ecology Limited. It had given him a basic idea of what he would see on the islands and the structure of the project. As far as he could tell, it was not much of an environmental impact statement, and he anticipated having trouble reaching conclusions about the ecological implications. There was just not enough known about the demographics of the turtle population on the northern Barrier Reef to assess whether the project was likely to be beneficial to their conservation.

When he got there, he saw that the islanders were raising both green and hawksbill turtles. The green turtle meat was intended for export as well as for local consumption. The plan was to sell the juvenile hawksbills as stuffed and mounted curios and to supply the current commercial demand that was fueling an unregulated fishery. Carr was troubled to see that Bustard had adopted the same dreamy-eyed plan of putting turtle poachers out of business by supplying a "superior product."

FIGURE 30 Visitors at Torres Strait island turtle farm, 1980.
Collection of the National Archives of Australia, AK21/2/80/47.

When it came time to write the report, Carr made it clear that the background document he had been given was inadequate and "not commensurate with the complexity of the enterprise proposed." As a consequence, his report was longer and more detailed than its sponsors may have intended, but this was necessary. Carr and Main went even further by including in an appendix an outline of what information a proper ecological impact statement would need to contain to make a realistic evaluation of the probable ecological impacts.[10] Carr was particularly dismayed that no studies had been conducted of the turtles that were to be cultured prior to the project's commencement, other than Bustard's initial nesting beach reconnaissance. There should have been a systematic census of each population, its distribution, and its seasonal movements through tagging at the breeding sites. Without this information it was impossible to judge how many eggs could be "harvested" for the farms without damaging the colony or to determine whether "head-starting" 10 percent of each lot of eggs collected and releasing them as yearlings was adequate compensation.[11]

Although it was beyond the terms of the inquiry, Carr and Main included a lengthy discussion of whether the world conservation organizations were likely to accept the project, a prerequisite for selling its products on the international market. In their opinion the prospects were not good. As they wrote this section, Carr reflected upon his recent correspondence about the turtle farm on Grand Cayman, and his thoughts crystallized his reasons for discounting its claims that farming is beneficial to turtle conservation. The principal benefits claimed were through augmenting the population through head-starting and then flooding the market and displacing poaching. Given the turtles' complex life cycle and long migrations, it was not possible to conclude that pen-reared yearlings would successfully recruit to the adult population until tag-and-release experiments could test this proposition statistically.

As for the market-flooding and price-lowering contributions to conservation, Carr and Main noted:

> Critics are predisposed to look with suspicion on all claims of this kind. There appear to be good reasons to doubt that markets for high-quality turtle products can readily be flooded. . . . While species-survival people are willing to give prospective farmers of wild species the benefit of the doubt, in the case of sea turtles they have been predisposed against commercial farming by Mariculture Ltd., which after promising to flood markets, is now doing just the reverse—purveying turtle meat through American supermarkets to housewives who never saw it before; and injecting new vigor into previously undersupplied markets for oil, calipee, skins and green turtle shells. In other words, the world's only currently operating turtle farm is actively expanding its market, raising prices and exacerbating the predicament of wild sea turtle stocks. If the Torres Strait project is to avoid a poor reception by conservationists, it will somehow have to inspire confidence in its ability to avoid such developments as mentioned above.[12]

To avoid the opposition of conservationists, Carr and Main counseled that five steps be taken:

1. present the project not primarily as a commercial venture but as one designed to maintain the economic and cultural integrity of a native people by placing most of the management in the hands of Torres Strait Islanders;
2. begin a "crawl breeding" program for mature turtles to produce all eggs to be used on the farm and make public a firm schedule for its development;
3. conduct and publicize a vigorous program for artificial selection of truly domestic breeds of turtles, with firm deadlines;

4. domesticate a strain of sea turtle that is bred not only for the desirable economic characteristics but that is also clearly distinguishable from and cannot be confused with wild-bred stocks when encountered either in the market or in the wild; and
5. make sure that all eggs for the farm come from the domesticated turtles and publicize the schedule for achieving this aim.[13]

Because their recommendations were so detailed, Carr and Main may have appeared to suggest that with quite a bit of fine-tuning the project could meet at least some of its objectives. However, their cover letter to Senator Willesee of the Australian Parliament conveyed a different message and makes clear they were counseling against the project. They cited two main reasons. First, the islanders' nutrition would decline. Once the green turtles were worth money and could be exchanged for cash, they would stop eating green turtles. That had happened to the Miskito of eastern Nicaragua when the turtle freezing plants were built and began exporting thousands of green turtles. Carr and Main's second reason was the risk of bringing so many outsiders to the islands, including tourists, business people looking for products, and even scientists eager to learn more about the islands' hawksbill population. This influx could have negative side effects unless visitors were very carefully restricted.[14]

Call in the Lawyers

When the directors of Mariculture learned that the US Fish and Wildlife Service had proposed to list the green turtle as endangered at the end of December 1973, they began to worry. As long as the green turtle was listed on CITES Appendix II, international law would allow its trade. The Cayman Islands government would be sure to issue CITES export permits, and Mariculture could still be in business.[15] But if the green turtle were classified as endangered under US law, the Interior Department might propose moving the green turtle from Appendix II to Appendix I at a future meeting of the CITES parties. Products made from Appendix I species could not be exported for commerce unless they were "bred in captivity." Would that phrase apply and allow the company to sell green turtle products in the United States? The directors knew that ten countries had to ratify CITES before it became legally binding, and they hoped that would happen by April 1, 1974. Could they argue that the CITES language somehow trumped the US proposed listing of the green turtle as endangered?

Although legal advice was getting expensive, the directors knew they would have to reassure their investors that the legal environment was still advantageous for the business plan. And new investors were in short supply. So Irv Naylor, Mariculture's president, decided to hire a Washington, DC, law firm to represent his company's interests before the federal agencies.[16] The law firm could make sure the federal law and the international treaty were applied consistently and, of course, favorably.

The company's new lawyers explained to Naylor that CITES would not in fact trump an endangered listing under US law. The treaty allows nations that are parties to the treaty to adopt stricter domestic measures—even to completely prohibit trade, possession, or transportation of any species.[17] But the lawyers had some good news, too. President Nixon had signed new endangered species legislation at the very end of 1973 to provide the legal basis for implementing CITES. The new law had a second category of species called "threatened." Conservationists liked this provision because it allowed the Interior Department to list a species before it was in extreme danger of extinction. But the "threatened" provision was also good for commercial operations. When a species was listed as threatened instead of endangered, there was no automatic ban on possession or imports; the Interior Department could write special rules for threatened species allowing some commercial use.[18] Being listed as threatened under the Endangered Species Act of 1973 was more akin to being on CITES Appendix II. Specimens and products from those species could be traded if the exporting country issued a permit saying the trade was not detrimental.

Naylor asked if it was too late to use the threatened provision now that the Fish and Wildlife Service had proposed the green turtle as endangered. The lawyers had other good news. President Nixon had signed the new endangered species law on December 28, 1973, the same day that FWS had proposed the endangered listing. They believed they could make a convincing argument that any listings which had been proposed under the superseded 1969 law had to be withdrawn and reproposed following the 1973 law's procedures and criteria. When Naylor looked puzzled, the lawyers acknowledged that the argument was a bit arcane but assured him that this was the kind of argument that appealed to the administrative agencies in Washington. Law firms who represented other commercial wildlife interests were going to make this argument, and it would probably persuade at least the US Commerce Department, if not the Interior Department.

The lawyers explained that Interior now shared responsibilities for the Endangered Species Act with the Commerce Department and that the Commerce Department's National Marine Fisheries Service (NMFS) was more accustomed to promoting commercial fisheries than regulating them. Before President Nixon had reorganized the federal government in 1970, the Fisheries Service was the Bureau of Commercial Fisheries, the bureaucratic descendant of the old US Fish Commission that much preferred restocking fish to regulating fishermen. NMFS was quite enthusiastic in principle about sea farming, and these were the folks who were going to be in charge of listings any marine species.[19] The lawyers' strategy for Mariculture, therefore, would be to make the case that sea turtles were under the jurisdiction of NMFS and that the Interior Department could not list the green turtle as endangered or threatened without its concurrence. A lot of fisheries trade associations would like to see sea turtles kept off the endangered species list because so many turtles were accidentally drowning in shrimp trawls. Mariculture, Ltd., could piggyback on their efforts.

Naylor knew that the legal fees for this strategy could be high, and Mariculture did not have a lot of money at the moment, but the US market was crucial to its success. Shipping through Miami to markets in Europe was also essential. He told the lawyers to give it a shot.

In February 1974, the director of NMFS told a lobbyist working for Mariculture that the green turtle's proposed endangered listing was a dead letter. Because the president had signed a new Endangered Species Act on the same day the listing was proposed, the new act superseded any species listings that were not already final. His agency now had shared responsibilities for marine turtles. If the green turtle was found to be "threatened" under the new process, by regulation NMFS could provide a permit system to allow importations of products such as those from Mariculture.[20]

When Wayne King heard about the letter, he realized the conservationists also needed a new strategy. Mariculture was not only seeking an exemption from the California faunal law; it was now lobbying to defeat a bill in the New York assembly adding sea turtles to the Mason Act. King would have to spend quite a bit of time in state capitols during the 1974 legislative session—and in every year that state legislatures were in session in Mariculture's target markets. However, a federal listing as endangered under the 1973 act would make these state-by-state rearguard actions unnecessary. States could give endangered species more protection if they were on the federal list, but not less. Therefore, King decided his best of course of action was to write a letter to the secretary of the interior asking the department to list the green turtle as endangered.

CHAPTER FOURTEEN

Conservation through Commerce

Wayne King was so busy working to defeat Mariculture's proposed amendment to the Mason Act, New York's faunal protection law, that he could spend only a bit of time on the petition to the US Department of Interior to list the green turtle. On April 23, 1974, he sent a letter to Secretary of Interior Rogers Morton, asking Interior to list the green turtle as endangered. He knew Interior's staff had data supporting the December 1973 proposed listing still in their files, including three reports Frank Lund had written on the status of foreign sea turtle populations. Rather than repeat all that material, King simply stated that it was the New York Zoological Society's scientific opinion that listing the green turtle as endangered was warranted. Renewed international demand for green turtle products, revitalized by a Caribbean turtle farm, was increasing an already too severe drain on wild green turtles. Moreover, the farm was promoting consumer demand by falsely claiming that turtle mariculture was an important asset in turtle conservation and could provide a new source of protein for starving nations. If Interior placed the green turtle on the federal endangered species list, its importation would be prohibited.[1]

Making full use of the new categories of risk, King asked also that the secretary list the loggerhead and Pacific ridley turtles as threatened. The loggerhead was being subjected to a variety of threats, including shoreline development adjacent to its nesting beaches in eastern Florida and ensnarement in the nets of shrimp trawling vessels operating offshore of Georgia and South Carolina. A threatened listing would provide a basis for selective implementation of a set of regulations to address these threats.

For supporting data, King referred Secretary Morton to information already on file with Interior's Office of Endangered Species and International Activities and to the reports of the 1969 and 1971 meetings of the

IUCN Marine Turtle Specialist Group. King concluded the letter by reminding the secretary that the United States had ratified the Convention on International Trade in Endangered Species of Wild Fauna and Flora, which listed the loggerhead and ridleys as Appendix II species. To conform to CITES, these two species had to be on the US list of threatened species. Although the green turtle was also on the Appendix II list, "the problem resulting from the proliferation of turtle farms has become critical since the Convention was drafted." This reason warranted listing the green turtle as endangered rather than threatened.

From King's perspective, if Mariculture succeeded in getting New York's Mason Act amended, it would be because of the same set of inflated claims the farm's supporters had made to the California assembly: the company's success in captive breeding and in augmenting wild stocks through head-starting. He would have to deflate these claims with data showing how few of the eggs laid at the farm actually had hatched. When the Mason Act amendment passed, King urged the governor of New York to veto it. Fortunately, the New York Department of Environmental Conservation made the same case, and Governor Wilson did veto the bill.

Mariculture did not take the New York governor's veto lying down. In October, the company's lawyers filed suit against both the environment and agriculture departments to stop them from enforcing the Mason Act against its sales in the state. They argued that the Mason Act did not apply to farm-raised green turtle products because they were not made from "wild animals." Moreover, importation of the farm's products would in fact serve the purposes of the act. Its products would greatly enhance the green turtle's chances of survival in the wild by displacing products made from wild-caught turtles. While it was true that Mariculture was still buying wild-laid eggs and hatchlings to build up its breeding stock, the lawyers argued that the products to be sold in New York were from turtles raised from doomed eggs and claimed that at least 1 percent of the eggs collected were released as yearlings on the nesting beaches.[2]

In January 1975, the trial judge granted Mariculture a preliminary injunction, barring the environment and agriculture commissioners from enforcing the Mason Act against its sale of marine turtle products in New York. The state attorney general appealed, and lawyers for King's New York Zoological Society filed a friend-of-the-court brief defending the state law.[3] King felt confident he could prove that the turtle products were from "wild animals"; the farm's own data showed its production depended on wild-caught eggs and adults. He had helped defend the Mason Act all the way to the Supreme Court when the crocodile shoemakers claimed it was unconstitutional.[4] This case, too, might take awhile, but he was confident that in the end his aggressive stewardship of Archie Carr's relics would prevail.

A Compromise in Sacramento

In the meantime, there was the matter of the California permits to deal with. Mariculture had succeeded in getting California's version of the Mason Act amended to allow companies to import marine turtle products into the state and market them if they obtained permits from the state's department of fish and game. If Mariculture could obtain a class I permit from the department, it could sell turtle oil to Polly Bergen's company, which in turn would need to get a class II permit to sell the face cream in California.

Commercial permits were generally not hard to get from the California department because, like most state fisheries agencies, it was favorably inclined toward commercial fishing interests.[5] But this particular permit was for marine turtle production, and the regulations required jumping through an unusually high hoop, thanks to a legislative compromise offered by John Hendrickson.[6] When representatives of the Sierra Club opposed the proposed turtle bill, Joop Schulz from the Suriname Forest Service had asked them to drop their opposition, explaining that Mariculture's purchase of eggs from Suriname's nesting beaches was crucial to his programs. Without this egg concession, there would be no money for any turtle conservation program in Suriname despite the significance of its rookeries to the Atlantic marine turtle populations.

John Hendrickson, who testified in support of the bill, had recommended a compromise, and the California legislature had accepted it. No business could get a permit to import farm-raised marine turtle products into the state if the department found that the import would adversely affect wild populations of green turtles. Before it could make this finding and issue a permit to Mariculture, the department had to consult the IUCN and be guided by its recommendations.

The California turtle law had a sunset clause, meaning that any import permit issued under it would be good only through the end of 1976. Then it would be necessary to go back to the legislature to get the law renewed. Luckily for Mariculture, one of the company's directors lived in Los Angeles and was happy to work with a lobbyist from the California Seafood Institute to get the permits and to renew the law when the time came.

Now that Mariculture needed an official endorsement from IUCN affirming that its production process would do no harm to wild turtle populations, its directors stepped up their efforts to get Archie Carr to take at least a neutral stance on the farm. They wrote directly to IUCN officials seeking their support and made a point to meet with Wayne King to describe the farm's breakthroughs in captive breeding and production that would soon reduce and then eliminate the need for wild-caught eggs.[7]

In addition to his job at the New York Zoological Society, King was now serving as conservation director to the Caribbean Conservation Corporation, the nonprofit organization supporting Carr's work in Costa Rica that had grown out of the Brotherhood of the Green Turtle. He was also now the vice-chair of the Survival Service Commission and therefore influential in IUCN circles. As chair of the commission's Crocodile Specialist Group, King was particularly interested in reviewing the data on Mariculture's captive reproduction and egg hatch rates. Captive breeding and commercial alligator farms had proliferated in the southern states ever since the 1970 endangered species listing of the American alligator, and King knew that hatch rates were critical to a farm's success.

When King met with Mariculture officers in New York City and reviewed the data that Mariculture believed supported their claim that the farm was achieving self-sufficiency in egg production, King thought the data showed the opposite. The farm was still collecting thousands of eggs from wild nesting beaches to maintain its production and would continue to do so far into the future if the trends in the hatching data continued.

From his meetings with Mariculture and his contacts in the Fish and Wildlife Service, King knew that the farm's Washington, DC, lawyers were preparing comments on his petition to list the green turtle as endangered. He assumed the lawyers would argue that since there were still thousands of *Chelonia mydas* in the world, the species could not seriously be considered endangered. Data on file with FWS and the green turtle's CITES listing on Appendix II suggested it was at most only a threatened species. The lawyers would probably challenge the lack of evidence supporting King's assertion that the farm's marketing activities were stimulating a resurgence of market hunting for green turtles. And to counter his claim that the farm was depleting wild rookeries to build its stock, they would argue that all the wild-laid eggs the farm obtained were authorized by appropriate government entities and were eggs that would not otherwise hatch.

On the off chance that the federal agencies might find these claims persuasive, King decided to submit a second petition, this one arguing that the green turtle should be listed as endangered under the similarity-of-appearance provision in the Endangered Species Act.[8] When the green turtle is raised on a commercial farm, its shell can be confused with the shell of the hawksbill turtle, the so-called tortoiseshell turtle, which was listed as endangered in 1970. He reckoned that the company's lawyers would have a hard time challenging this petition; the company's marketing brochures said the farm-raised shell was superior to wild-caught tortoiseshell. And he knew many FWS enforcement agents would corroborate his claim that they could not differentiate listed from unlisted sea turtle specimens once they were processed into products.[9]

To King's surprise, however, the law firm submitted not comments, but a petition on behalf of Mariculture, asking the government to list *Chelonia mydas* as threatened. The petition in fact said very little about the status of the green turtle. Most of its assertions were about the company and how it was improving the survival prospects of the green turtle by rearing and releasing a significant portion of the eggs collected. By 1979, a mere five years from the time of the petition, Mariculture would no longer need to take any wild eggs.[10] The lawyers were asking the secretaries to include an exemption for farm-raised products in the protective regulations that would accompany the green turtle's listing.

King suspected that the federal agencies had had advance warning of the farm's petition because, the day after it was submitted, they published a joint notice in the *Federal Register*, the daily publication of US federal agencies, announcing they were conducting a formal review of the status of green, loggerhead, and Pacific ridley turtles and requesting information that was relevant to this review.[11] In fact, Carleton Jones, the farm's Washington, DC, lawyer, knew that at least one of the federal agencies that would make the listing decision, the National Marine Fisheries Service (NMFS), was not eager to list any commercially exploited marine species.[12] Jones's strategy was to give the agencies an alternative course of action. Using the threatened category of the new ESA, the agency could fashion a special rule allowing the green turtle to be bred and raised in captivity and marketed in the United States indefinitely. For the period of time it would take the farm to perfect its production methods—say, for instance, a period of ten years—Jones suggested that the rules allow the sale of products made from wild-caught, doomed eggs. At the very least, this proposal would give the California department enough cover to issue Mariculture its class I permit.

Carleton Jones was confident that his regulatory proposal would appeal to the Fisheries Service but had doubts whether the staff in the Office of Endangered Species at FWS would buy it. At the CITES conference in early 1973, some members of the US delegation had been wary of the exception for Appendix I species that allowed trade in specimens that were bred in captivity for commercial purposes, thinking it created a potentially huge loophole for illegally poached wildlife products and an enforcement nightmare.[13] But Fish and Wildlife was beginning to get pressure from the southern states to reclassify the American alligator as a threatened species. This would allow skins from alligator farms to be sold in the United States and abroad under the CITES exception. Soon, they would have to make decisions interpreting the CITES provision, whether they wanted to or not. Jones knew that the agencies were discussing a draft memorandum of understanding on how they would share their responsibilities under the ESA.[14] With a little luck and some judicious lobbying, the fishing-friendly NMFS would be making the call on how the green turtle should be listed.[15] In the meantime, Mariculture pursued its application for a class I permit with the California Fish and Game Department.

A Task with Force

Because California's new turtle mariculture law required the fish and game commissioner to seek recommendations from the IUCN before issuing a permit, pressure was mounting on the marine turtle specialists to visit Mariculture and determine if it was indeed making progress toward its goal of total independence from eggs collected at wild turtle rookeries such as Suriname and Ascension Island. As co-chairman of the marine turtle group, Tom Harrisson knew that the opinion of the specialists was divided and consensus would be hard to reach. Besides, IUCN official policy on the issue of commercial use was unclear.

Harrisson's idea was for IUCN to convene a small task force on the commercial use of marine turtles that could hammer out a set of standards or principles against which the operations of Mariculture could be compared.[16] The task force would meet in Florida to develop its principles, and then a small group would go on to Grand Cayman to see if the farm met those standards. Archie Carr agreed with the plan and made recommendations on who should participate. He urged IUCN to appoint Wayne King to the task force, given his intimate knowledge of CITES and state faunal laws. He wanted David Ehrenfeld, Peter Pritchard, and Nicholas Mrosovsky to be there, and also George Balazs, who had information indicating that the farm was marketing its products in Hawaii. He agreed that Leo Brongersma should not be left out, and it seemed important to include Joop Schulz and George Hughes, with their perspectives from turtle conservation programs in Suriname and South Africa. But somehow, John Hendrickson, who had been the one to propose IUCN's involvement in the California permit matter, was not on Carr's list of specialists who should get an invitation from IUCN to join the task force.

David Ehrenfeld agreed to help draft the IUCN principles. This was an opportunity to base the standards on the economic and ecological realities of mariculture he had explained to Robert Schroeder at Tortuguero almost ten years before and he had elaborated on in his paper in *American Scientist*. But he would not take part in the site visit to Mariculture on Grand Cayman, finding it unethical under the circumstances to accept the farm's hospitality.

The statement of principles developed by the ad hoc task force of specialists condemned certain commercial uses of sea turtles, like selling stuffed yearlings as curios, but not turtle farming. Instead, the statement set out standards that commercial farms needed to meet to be consistent with conservation principles.[17] When the task force members who went to visit Mariculture met the farm's owners and staff, they did not give them the new statement of principles, since they had not yet become official IUCN policy. They intimated, however, that some task force members doubted whether the farm met the standards for conservation required by the California law.

A few months earlier, the farm's marketing staff had unwisely issued a publicity brochure and held a press conference in London touting their success in conquering the captive breeding conundrum. They claimed that the eminent British scientists on their team had solved the problem by introducing two wild males into the breeding pool, triggering a mating frenzy. The females were now laying eggs with abandon, and soon the resulting turtles would be ready for market.[18] The task force visitors told the farmers that they considered these claims to be inflated and possibly a sign of bad faith.

Antony Fisher agreed that the marketing staff had somewhat exaggerated their success. The new CEO, Johnny Johnson, a retired biochemist and experienced businessman, assured the visitors that these claims were no longer in circulation. With Sir Alan Parkes and Dr. Amoroso advising, the farm was operating on a more scientific footing. Parkes had in fact pulled together the breeding records and lab reports, which they were welcome to examine. Before the visiting specialists could meet with Parkes and Amoroso, Wayne King and Archie Carr left the farm for the airport. They had attended only one short meeting and taken a tour of the farm. Antony Fisher felt relieved that their harshest critic, Wayne King, had left but was chagrined that Carr would miss the discussions with his eminent scientific consultants.[19]

Despite his earlier trepidations, Alan Parkes thought the discussion of breeding performance went well.[20] The visitors knew a lot about the green turtle's behavioral ecology, but none had ever worked with a captive breeding population. He and Amoroso had "something like a walkover" in the discussions of behavioral cues and reproduction, incubation, and embryonic development. He showed them the draft of a paper and supporting data from the 1973 breeding season they would soon publish in the *Journal of Zoology.*

The most interesting part of the discussion came when Tom Harrisson brought up the question of sperm storage in the female turtle. Parkes knew from a previous meeting with Wayne King that this issue was central to King's claim that the farm did not have an effective captive breed-

FIGURE 31 Archie Carr on inspection tour of Mariculture, November 25, 1974, with members of the ad hoc task force after drafting IUCN Principles for Commercial Exploitation of Sea Turtles.
Courtesy of Peter C. H. Pritchard.

ing stock and would always require wild-laid eggs, contrary to the requirements of the California turtle mariculture law. The farm would thus further drain wild populations. King had argued that the farm's data showing a decline in the number of eggs that hatched strongly suggested that females were beginning to run out of sperm they had stored from their matings in the wild.

Parkes got on especially well with Tom Harrisson and thought that at the very least he had convinced Harrisson that Wayne King's sperm storage explanation for declining hatch rates was incorrect. Parkes said it was true that one female green turtle had laid several fertile clutches over a period of weeks after only one mating at the beginning of the season. So storage for a period of weeks or even months was likely. But in his dissections he had seen no evidence that females could store spermatozoa for years. And the fact that none of the captive wild females had laid eggs until after mating in the breeding pool started in April 1973 made the idea that they had stored sperm for several years after mating in the wild highly unlikely. He emphasized that once the farm-born and -reared females matured and started to lay fertile eggs, they would have conclusive proof. At the end of their discussions, Harrisson seemed less adamant in his opposition to the farm.

After the formal parts of the meeting were over, Parkes enjoyed getting to know George Hughes from Durban and Stanley de Silva from Sarawak, two members of the IUCN marine turtle group who had taken no part in the attack on Mariculture, and invited them to dinner with some of the farm staff. He had pleasant conversation also with Leo Brongersma—who told Parkes that although he considered himself a conservationist, he enjoyed his turtle soup—and with the Survival Service Commission's Tony Mence, with whom he shared an interest in the wildlife of east Africa. He was astonished, then, when he read the report Mence prepared, which IUCN's deputy director sent to Mariculture a few weeks later.[21]

The report said that the visiting panel considered the extent to which the company was in compliance with the ad hoc task force's principles and recommendations on farming. Due to the company's misleading and untrue statements, an unjustifiably favorable impression existed in some quarters and grave suspicion of its integrity in others. It was clear that "reconciliation of the conservation ethic with business efficiency in the company's policy" remained to be demonstrated. As no long-term operational planning projection had been made, the viability of the farming operation had yet to be proven. "For these reasons alone the panel is unable to endorse the operations of Mariculture Ltd. as making a positive contribution to the conservation of the green turtle." Parkes was particularly astonished by this statement. There was not even a grudging acknowledgment of their having achieved the first-recorded captive breeding of the green turtle, which surely would prove to be a breakthrough for its conservation. Why was it so hard for the IUCN panel members to "distinguish between the shortcomings of Mariculture and the importance of its results?"

The IUCN panel had nevertheless commended the company for its willingness to withdraw the inaccurate publications, to clear future ones with its scientific advisors, and to maintain frank and open dialogue with IUCN in order to ensure compliance with conservation requirements. The panel noted that "valuable research programs, some of outstanding scientific merit, are in varying stages of development, and that facilities are also made available by the company to private research workers." Although "the present operations could not be endorsed as being in the conservation interests of the Green turtle," IUCN was ready to endorse the company's operations when convinced that they conformed to conservation principles, "recognizing that new activities or significant extensions of present ones . . . without previous notice will be regarded as a breach of such assurances and therefore as grounds to doubt the company's integrity of purpose."

Immediately following the task force meeting, Wayne King had sent a letter to California's director of fish and game, Charles Fullerton, alerting him that IUCN was reviewing a set of principles for commercial exploitation prepared by a special task force of marine turtle specialists. He suggested that a permit from the department for Mariculture might not meet these principles.[22] When Leo Brongersma learned of King's precipitous letter concluding that Mariculture was not operating in conformity with the agreed-upon principles, he took King to task for it. In a letter to King that was also copied to Tony Mence, Brongersma said: "I am astonished by the strong bias and vindictiveness shown by some of the opponents to turtle ranching and farming, and in this I see a proof of the weakness of their argument."[23]

But Fullerton had also received a copy of a letter that Henk Reichart, a resident of California, had written to his state senator. Working closely with Joop Schulz in Suriname, Reichart had started a small turtle farm at Bigisanti Beach with the help of Mariculture. He warned the senator that the task force had not gone to Grand Cayman with an open mind.[24]

Five months later, Fullerton received a letter from Frank G. Nicholls, IUCN's deputy general, stating that on the basis of the specialists' review, the executive board had concluded that no existing large-scale commercial turtle culture operation currently met IUCN's Principles and Recommendations on Commercial Exploitation of Sea Turtles. Fullerton believed that the specialists had indeed gone to the farm with their minds made up, but without a positive recommendation he could not issue the permit.[25]

When the managers of Mariculture learned they would not receive IUCN's endorsement for the California permit, they were so close to declaring bankruptcy that this information might not have even registered.[26] A bank that had agreed to invest in the farm had just gone bankrupt, and the farm's other underwriters were not willing to extend more credit. Relations with the Commonwealth Development Finance Company, the company's chief underwriter and investor, had frayed badly as the farm built costly expansions, made plans for a second farm four times the size of the Mariculture's principal facility at Goat Rock on Grand Cayman, and prepared to sell franchises around the Caribbean. With the bills mounting and full-scale production still months away, on May 15, 1975, the farm's principal creditor, the First National City Bank of New York, placed Mariculture in receivership. Robert Moyle, the Price-Waterhouse accountant who worked for the bank and was now in charge of Mariculture, began looking for a buyer who was interested in raising sea turtles.[27] Five days later, NMFS and FWS jointly published proposed rules listing the green turtle as "threatened."[28]

A London Bridge

Mariculture's lawyer, Carleton Jones, called Moyle from Washington, DC, to explain what the federal agencies had decided.[29] The proposed regulations prohibited anyone from importing green turtle products into the United States, and within one year it would be illegal to ship them in interstate commerce. But there was one important exception: green turtles or their products could be shipped and sold if they were derived from mariculture—that is, if they were "bred *or* raised in captivity."[30] To get a permit under these rules, Mariculture and its customers would have to demonstrate that the green turtle products were "taken for or derived from a captive population that is demonstrably in the process of becoming a captive breeding population that is completely self-sustaining and independent of wild stocks, but is temporarily sustained in part by the addition of turtles or eggs taken in the wild, the taking of which is demonstrably not a major threat to wild stocks." After two years, the Services could not renew the annual permit unless the permit holder could demonstrate that their products were "derived from a closed-cycle farming operation consisting of a captive-bred population which is completely self-sustaining and independent of wild stocks."[31]

Jones told Moyle he realized two years was not going to be enough time for the farm to "close the cycle," so to speak. The turtles that hatched after the 1973 breeding frenzy were less than two years old; it would be at least five more years before they could be considered breeding livestock—five years of feeding them the expensive turtle chow and keeping them healthy. Moyle told him these terms would be impossible to meet; no buyer in the world would agree to feed stock for that long with no hope of a return. The agencies might as well not include a mariculture exemption. Jones said he would submit comments on the proposal asking that there be either no time limitation or that the agencies give the farm at least five years to meet the self-sufficiency criterion. He would repeat what he said in his 1974 petition: mariculture benefited the green turtle species by increasing knowledge of its biology and, ultimately, by reducing pressure on wild populations.

For the next several months, Robert Moyle advertised the farm with an asking price of $2.5 million. Before he had any hope of finding a buyer, he needed to remove the environmentalists' objections to the operation. He read and reread the IUCN principles and discussed them with Jim Wood. The former graduate student of John Hendrickson was now the farm's research manager. Was there any way that *all* the marine turtle specialists could agree that the farm would ultimately serve "conservation through commerce"?[32] Were their objections based on scientific and technical matters or philosophical principles? The only way to find out was to ask.[33]

In August 1975 Robert Moyle contacted IUCN in Morges and requested a meeting with members of the specialist group in London. The IUCN executive body had just adopted the marine turtle group's principles and recommendations as official IUCN policy, and it seemed appropriate to find out whether IUCN thought Mariculture could ever conform to them. It seemed to Moyle that the major concern was that the farm would increase demand for turtle products and this would in turn increase the illegal killing of wild turtles.

When the meeting took place, Mariculture's CEO, Johnny Johnson, was pleased to see that the farm's main opponents—Archie Carr, Wayne

King, and Tom Harrisson—had decided to attend. He was also happy that Joop Schulz, from the Suriname Forest Service, was there. The nesting beaches under Schulz's management were the farm's main source of wild-laid eggs, and Jim Wood's projections showed they would still need these eggs a little longer until the breeding stock was working at full capacity. That the turtle specialists respected Schulz for the scientific rigor of his conservation program was evident from the comments of Leo Brong-ersma, his fellow Dutchman. Johnson could tell from the way the others spoke of Schulz's program that they agreed. The IUCN official in atten-dance was Tony Mence, and he seemed eager for the specialists to reach some understanding with the farm. Mence said IUCN was not in princi-ple opposed to culture operations as long as they were subject to reasonable conditions such as those the task force had recommended. He seemed satisfied with Moyle's assurances that Mariculture had abandoned its expansion plans and with Wood's new operational plan, which projected no further need to take wild turtles or eggs after 1979.

When it was over, Johnson thought the meeting had been productive but was not sure where things stood with IUCN. He was therefore quite pleased to receive within a month of the meeting a copy of a letter that IUCN's Frank Nicholls wrote to Charles Fullerton at the California De-partment of Fish and Game. Nicholls said that a panel of IUCN experts had undertaken an additional review of Mariculture since it had been placed under receivership and that a new operational plan had been devised. Nicholls wrote:

> Although IUCN retains some reservations regarding the feasibility
> of this plan, it nevertheless recognized that a reasonable effort had
> been made to formulate a management regime consistent with the
> spirit of IUCN's "Principles", and that pledges of good intent had
> been given. IUCN is therefore prepared to accept this as earnest of
> the new management's good faith and for as long as such good faith
> remains demonstrated and the provisions of the above mentioned
> operational plan adhered to, IUCN will raise no objection to the
> issuance of a Class 1 or Class 2 permit to the company.[34]

Out in Sacramento, Fullerton now had the go-ahead he needed to issue permits so that Mariculture could sell turtle oil and other products in

California. This would greatly improve the company's prospects of finding a buyer. Fullerton was glad finally to be able to do so.[35] He had toured the farm in 1975 and thought it was just like a fish hatchery. He was pleased to know that the "feds" had followed California's example and made provision for a mariculture permit in their proposed sea turtle regulations. As he had written in the department's comments to the two federal agencies on the proposed regulations, "We believe the laws and regulations developed in California will materially aid in encouraging rehabilitation of the world's marine turtle populations while realistically allowing some legitimate use of the resource. We have enclosed copies of our laws and regulations for your review. We would be pleased to assist you in any way we can to help perpetuate a positive marine turtle program."[36]

By November 1975 Moyle thought he had a potential buyer. Heinz Mittag was a German businessman who had made a small fortune when he sold his personal hygiene products company to Johnson & Johnson, the giant consumer products corporation. Mittag visited the farm while on Grand Cayman to meet his accountant. The Cayman government officials were pleased to give him a tour of the farm, and Mittag felt pity when they told him the turtles would be slaughtered if a buyer was not found. He was intrigued by the prospects of raising food from the sea, helping to alleviate food shortages while at the same time helping to conserve an amazing sea creature. His wife, Dr. Judith Mittag, the gynecologist who invented the OB tampon that had made their fortune, was even more fascinated by the turtles and confident they could sort out their reproductive problems. Then, after reading Archie Carr's books, she was anxious to help the cause of turtle conservation.[37]

After hearing the saga of the California permits, the Mittags were not naïve about the problems they would have with the American environmentalists. Before they would commit to the purchase, they sought assurances that Frank Nicholls' letter reflected the views of IUCN's marine turtle specialists. When the Mittags met with Tony Mence in Dusseldorf, he assured them that as long as they maintained the new operational plan, with its goal of gaining independence of wild-laid eggs, and did not expand into new markets without consulting IUCN, there would be no major objections. He recommended a name change to signal that the farm was under new management. The name, Cayman Turtle Farm, Ltd., was suggested and agreed to. It conveyed the Cayman Islands' heritage of providing sea turtles to the world. Although the operation was not yet a farm in the true sense of the word, it would be one soon, once all the technicalities were ironed out.[38]

CHAPTER FIFTEEN

The Best Available Science

With the California turtle mariculture permit in hand and an enthusiastic buyer, Cayman Turtle Farm needed only to get the federal permit. But the final federal regulations authorizing green turtle mariculture were nowhere to be seen. The Fish and Wildlife Service and the National Marine Fisheries Service could not agree whether the green turtle was endangered or threatened. Without this agreement, there could be no federal regulations and permits for mariculture.

At first, the agencies agreed to leave sea turtles out of the memorandum of understanding on the Endangered Species Act they signed in August 1974. Each had thought they should take the lead on making decisions regarding sea turtles: whether and how turtles should be listed and if regulations should allow them to be taken for subsistence or accidentally in commercial fisheries.[1] Rather than flip a coin, the agencies agreed to try to decide these policy issues jointly. They had three petitions to respond to, urging them variously to list the green as endangered, as threatened with an exception for mariculture, or as "threatened due to a similarity of appearance" to the endangered hawksbill.

By July 1975, NMFS was insisting that the green, loggerhead, and Pacific ridley turtles be listed as threatened. This would give the service the flexibility to fashion regulations that allowed the species to be taken, imported, farmed, or otherwise affected by human activities—an approach more consistent with its managerial approach to living marine resources. The endangered species staff at FWS thought the data indicated that green and Pacific ridley turtles were endangered by the heavy exploitation taking place in several regions and that these regions constituted a significant portion of the species' range. To make this case, the staff was digging into the available data to fine-tune the listing at the population level, which the ESA's definition of "species" allowed them to do. The FWS staff were flabbergasted when, out of the blue, came word that the NMFS endangered species program had decided to write an environmental impact statement. NMFS said that the amount of public comment it had received on the proposed threatened listing indicated that the action was clearly controversial; it was thus appropriate to consider the impacts of the marine turtles listing in an environmental impact statement. An EIS would entail a public hearing to take additional comments on the range of questions the proposed listing raised.

To some of the FWS staff, this decision looked like a transparent attempt to delay the regulations, to give Cayman Turtle Farm more time to become self-sufficient in egg production. But NMFS officials insisted that the EIS and the additional public comment period would help the agency decide whether to afford an exception for mariculture. If the decision later was to dispense with the exception, the EIS would help defend the decision if, as seemed likely, the farm's lawyers decided to sue.

The FWS staff did not buy that reasoning. Their director informed the NMFS director of his strong opposition to the proposed delay: "We cannot agree with delaying this listing action for the possible benefit of a commercially-oriented organization which already has had ample opportunity to submit its comments in writing. . . . We also oppose any delay for purposes of preparing an environmental impact statement. The existing impact assessment is biologically sound and covers the situation in adequate detail. It was jointly prepared by professionals in both our organizations and I feel their product clearly supports a negative declaration [on the need for a full EIS]."[2]

A Tactical Statement

When the FWS staff received the draft EIS, they realized the text was taken almost verbatim from materials that FWS had prepared. That NMFS had not added new information to the analysis fueled their suspicion that the EIS was a delaying tactic. The Interior Department lawyers advised them there was little recourse for this breach of their interagency agreement; the secretaries would have to work out their differences.[3]

After one postponement to allow NMFS to finish the EIS, the day arrived for the hearing on February 25, 1976.[4] Because so many people had asked to speak, NMFS held the hearing in the penthouse conference room in Page Building 1, one of the twin office buildings occupied by

NMFS's parent agency, the National Oceanic and Atmospheric Administration, in the Georgetown section of Washington. A lawyer from NOAA's Office of General Counsel, Paul Kiefer, was the presiding officer. He asked people to address first whether the agency should list the sea turtles as threatened or endangered and then to comment on whether the special rules should allow any turtles to be taken in the course of commercial fishing, for subsistence use, or for mariculture.

Several fishery representatives spoke in opposition to the listings if they meant that fishermen would be required to take expensive steps to avoid drowning sea turtles in their nets. Others asked for an expansion of the exemption for subsistence use for the Pacific island territories. But by far the largest amount of testimony concerned the mariculture exemption. There were now two opposing camps, on either side of the issue, and each camp had a team of lawyers who presented voluminous factual assertions and legal arguments to support their side.

As this was NMFS's first major decision under the Endangered Species Act of 1973, it was under close public scrutiny. Kiefer was fairly certain the agency would be sued by one interest group or another. If the agencies "split the baby" in this decision, it was conceivable they would be sued by both sides. As he struggled to keep order in the meeting room, Kiefer tried to guess which of the lawyers would be the one to sue his agency, and whether the Justice Department would let his attorneys assist in the defense. If he had to bet, he thought it would be the shrimp fishermen who would sue.[5] They believed they should have a complete exemption from any turtle protection regulations. They insisted that shrimp trawls caught very few sea turtles; if sea turtles were threatened with extinction, it was more likely that coastal development and pollution were the culprits.

Wayne King spoke on behalf of the New York Zoological Society and a host of conservation organizations.[6] He presented lengthy testimony, drawing on a forty-seven-page analysis of the EIS and the proposed regulations. He had several major points. First, it was wrong for NMFS to try to delay the listing so it could weaken the proposed regulations. Given the inadequacies of the EIS, it was clearly appropriate for FWS to have sole jurisdiction over sea turtles. NMFS had given no indication of where shrimpers could fish to avoid catching sea turtles because it had no idea of where the turtles' breeding, foraging, and nesting areas were and no plan to identify this "critical habitat" even though this was required by the ESA. Nor had the agency required the shrimp fleet to develop and use a net that would exclude turtles rather than catch and drown them.

King heaped the greatest criticism on the proposal to include a two-year permit exception for mariculture. Again, he argued, NMFS's EIS showed that the agency understood very little about conserving sea turtles and should not have jurisdiction. Despite the company's new projections, there was no hope that Cayman Turtle Farm would attain self-sufficiency from wild-caught eggs in that or any other period of time. Its own data made clear that breeding on the farm was ineffective and that the farm was relying on stored sperm to fertilize the few eggs that were being laid. It would never become closed-cycle but would always depend on wild-laid eggs. This would eventually deplete the Suriname turtle rookery, one of the most significant in the entire Caribbean. By marketing sea turtle products around the world, the farm would encourage others to take turtles illegally in order to cash in on this demand. Poaching was very difficult to prevent, given the remote locations of sea turtle nesting beaches. As IUCN had found in 1975—basing its conclusions on the principles and recommendations of the ad hoc task force—turtle farming was not in the conservation interests of the green turtle.

The lawyers for the environmental groups agreed with King's assessment and added several points of their own. The mariculture exception would make it "extremely difficult if not impossible to ensure that any by-product available in the United States is indeed from a farmed turtle. Thus, wild turtles will still be taken and their by-products imported illegally into the United States under the guise of farmed by-products."[7] They urged instead that all importation and sale of by-products from the three species be banned immediately by listing the three as endangered, a classification warranted by the available biological information. If captive breeding of sea turtles proved necessary for their conservation, the ESA allowed "captive, self-sustaining populations" of otherwise endangered wildlife to be classified as threatened. However, if FWS and NMFS deemed it appropriate to classify the three species as threatened, no exemption should be allowed for mariculture that permitted the further taking of wild turtles or their eggs until it could be shown that mariculture operations would enhance the species' survival prospects.

William Butler, lawyer for the Environmental Defense Fund, summarized their arguments: "The miniscule adverse domestic economic impact that would ensue from listing these three species as endangered, the frightening rapidity with which their populations are declining, and the excessive length of time taken for their listing (even compared with other sea turtles) all militate towards fast action now before it is too late." He said he was appalled to learn from the draft EIS that the US Agency for International Development had financed one or more of the Nicaraguan turtle processing facilities that began operations in 1969 and 1970. As part of the listing action, USAID should "be informed as to the undesirability and potential illegality of such action."[8]

Carleton Jones, who had drafted the listing petition for Mariculture eighteen months before, made the case now for Cayman Turtle Farm. He argued that the green turtle should be listed as threatened and farm-raised products exempted from the import prohibition, but without the two-year time limit in which to close the farm stock's life cycle. This was an arbitrary and unreasonable limit, he argued, in view of the absence of any evidence that wild turtle populations had been harmed by the moderate numbers of eggs Cayman Turtle Farm purchased from the Suriname rookery, eggs that would otherwise have been sold in local markets to support the hatchery program. In any event, the farm was projecting that it would achieve self-sufficient production by 1979 or 1980, and it was reasonable to extend the exemption until that time. Jones submitted a statement for the record by the farm's reproductive biology expert, Sir Alan Parkes, that the breeding problems of green turtles in captivity would soon be solved. Parkes and Dr. Amoroso had several lines of investigation under way that would explain the declining hatchability rates of the farm-laid eggs. Depletion of stored sperm was not one, but was in fact the least plausible explanation. It could be the high density of breeders in the pool or the absence of an essential amino acid in their diet.[9]

Jones stressed that all the criticisms that Wayne King had levied regarding false claims and marketing were no longer applicable because the farm was under new ownership and management. There was no plan to expand the market for turtle products. In fact, the marketing and operational plans were fully consistent with IUCN's Principles and Recommendations Concerning the Commercial Utilization of Marine Turtles and with NMFS's plan to expand aquaculture production.

Finally, the inclusion of a mariculture exemption would be consistent with the CITES provision allowing trade in by-products of specimens that are bred in captivity for commercial purposes.

A Turf Battle in Berne

While NMFS took public comment on the EIS, the evidence was mounting that exploitation rates for the Pacific ridley and green turtles were astronomical. The endangered species staff at FWS wanted the three unlisted sea turtle species listed quickly, before the upcoming conference of the parties to CITES, scheduled for November 1976 in Berne, Switzerland. Some of the FWS staff thought they should concede primary authority to list sea turtles to NMFS as long as it would agree to list the green, loggerhead, and Pacific ridley as threatened immediately under the "similarity of appearance" provision of the ESA.

Staff in the policy office in the Department of Interior, however, vetoed the proposed compromise and pushed FWS to pursue the "similarity of appearance" listing. They told FWS to argue that because NMFS had decided unilaterally to prepare an EIS, the final listing decision would be delayed until all the comments could be analyzed and the final impact statement produced. Products made from the three species could not be distinguished from those made from the hawksbill and Atlantic ridley turtles, which were already listed as endangered. By the time the final EIS was completed, thousands of these turtles might be taken and labeled as some other species.[10] NMFS found this argument persuasive, so in June, FWS and NMFS published proposed rules listing the three species as threatened and banning their import, sale, transportation, and possession, except under permit. Once the NMFS EIS was final, the primary listing regulations would supersede these rules.[11]

The two agencies then met to prepare a position paper on the sea turtle proposals that would be debated at the first conference of the parties to the CITES treaty in Berne. A proposal by the United Kingdom asserted that there was sufficient data on international trade to warrant uplisting to Appendix I the loggerhead, the green, and the Pacific ridley.[12] Before the US delegates flew to Switzerland, the US agencies agreed that the United States would oppose this move on grounds that the data did not support it. Having only recently proposed listing all three species as threatened, the secretaries of interior and commerce were on record that the three were not yet in any immediate danger of extinction. But the FWS staff had never been happy with that classification; some turtle populations were very clearly at risk of extinction. So the US delegation, composed of representatives from both agencies and their various advisors, went to the first conference of the parties to CITES with dissension in the ranks. But everyone on the delegation agreed it was important for the parties to adopt guidelines on the criteria for adding or removing species on Appendix I and II.

At the CITES conference in Berne, the most contentious issue turned out to be the uplisting of the remaining species in the family *Chelonidae*. The West German delegation strongly opposed moving the green turtle to Appendix I, thus banning international trade, and demanded a formal vote. Germany was the home of one of the largest buyers in Europe of green turtle meat and calipee, the soup maker Lacroix. Delegates from Ghana and Zaire had threatened to walk out if Germany's opposition prevailed. These nations were frustrated that the green turtle populations of western Africa had been shipped off to Germany for the soup industry. When the proposal was debated in committee before the plenary session, Lynn Greenwalt, director of FWS and the senior US delegate in the room, did not forcefully oppose the uplisting despite the preconference agreement that the US position would be to oppose. He merely said that if the proposal passed and the green turtle was added to Appendix I, the United States would adopt a liberal interpretation of the "bred in captivity" exception, to treat green turtles from mariculture as Appendix II species, allowing turtle farms to prove that they could become self-sustaining.[13]

When Germany insisted on a formal vote on the green turtle's transfer to Appendix I, Curtis Bohlen, the head of the US delegation, abstained instead of voting. He did so even after Prudence Fox, NOAA's head of international affairs and vice-chair of the conference, came down from the head table to ask him to vote no. Bohlen was a senior assistant to the secretary of interior, Rogers Morton, and an old hand in conservation diplomacy. As he held his tongue, he could see the delegation's scientific advisor, Wayne King, sitting nearby. The proposal to uplist all *Chelonidae* to Appendix I except the flatbacks and green turtles of Australia was approved by a vote of fourteen to five, meeting the required two-thirds majority vote.[14] West Germany later filed a reservation on this change, allowing it to continue importing green turtles.[15]

When he learned that the uplisting proposal had passed, Robert Stevens, NMFS's endangered species administrator, worried that FWS now might pressure NMFS to change their jointly proposed listing from threatened to endangered to conform to the decision of the CITES parties. When he returned from Berne, he briefed the fisheries management office at NMFS on the potential ramifications of the CITES decision:

> The direct effect would be minimal, provided the present maricul-
> ture operation is allowed to continue on an open-cycle basis for two
> years in order to conform with our proposed domestic mariculture
> exception. The impact on incidental catch by shrimpers would be
> nil, because the Convention is not concerned per se with take but,
> rather, with trade.
>
> The indirect effect could be catastrophic depending on whether
> the United States now feels obligated to list the three turtles on our
> endangered list in order to conform with the action by the parties to
> place them on Appendix I. Strictly interpreted, endangered species
> under the ESA of 1973 can neither be legally taken by shrimpers nor
> reared nor imported for commercial utilization. This proposition
> along with the question of continued joint jurisdiction has now
> intruded upon the rulemaking process and our long frustrated
> efforts to list the green, loggerhead, and Pacific ridley on the
> threatened species list.[16]

The decision on how to classify the green turtle was now an interagency turf battle of major proportions.

A Final Decision

Pressure was mounting to make the final listing decision. Thanks to the EIS hearing, NMFS now had volumes of public comment opposing and favoring listing the green turtle as threatened and as endangered and arguments both for and against a mariculture exemption.[17] The CITES parties had uplisted all species in the family *Cheloniidae* to Appendix I, a decision FWS had tacitly supported, and FWS had promptly issued regulations to implement CITES.[18] It was now 1977, two years since the listing as threatened had been proposed, along with special rules on subsistence, fishing, and mariculture. It was time to find a compromise.

The agencies decided to list as endangered the sea turtle populations that were the most depleted or were suffering the highest rates of exploitation. These were the Florida and Mexican green turtle nesting populations and the Mexican populations of Pacific ridleys. All other species in the family *Cheloniidae* were listed as threatened. Once these classifications were agreed to, they could settle on which agency had jurisdiction for future policy decisions. After weeks of back and forth discussions, a solution was reached. Lynn Greenwalt for FWS and Robert Schoning for NMFS signed a document entitled "Memorandum of Understanding Defining the Role of the U.S. Fish and Wildlife Service and the National Marine Fisheries Service in Joint Administration of the Endangered Species Act of 1973 as to Marine Turtles."[19]

Under the agreement, NMFS would have sole jurisdiction over "sea turtles, including parts and products, when in the marine environment and activities impacting turtles in the marine environment. This includes the waters adjacent to sea turtle nesting beaches." NMFS would make the call on how to restrict fishing activities that involved encounters with sea turtles; FWS would have sole jurisdiction "over sea turtles, including parts and products, when on land." Because sea turtles spend so little of their very long lives on land, this arrangement left some people scratching their heads, everyone except those who knew about the controversial turtle farm. Under this language, it seemed that FWS would make the policy decision on mariculture. When the Senate Environment and Public Works Committee held oversight hearings on the ESA, Senator John Culver asked NMFS deputy director Jack Gehringer what would happen if they found a turtle that could fly. Culver proposed that jurisdiction should go to NASA.[20]

It took NMFS another year to finish the final EIS, to summarize all the comments, and to describe the rationale for the final decision.[21] The two services published the final protective regulations on July 28, 1978. When the managers of Cayman Turtle Farm received the news, they were astonished to learn there would be no exception made for mariculture at all, not for two years nor for any other period of time. In a little over thirty days, it would be illegal to sell their turtle products in the United States or ship them through Miami to their customers in Europe.[22]

There would be an exception made, however, for incidental takes of sea turtles in commercial fisheries and for the traditional level of subsistence turtle harvests in the US Trust Territory of the Pacific. The EIS explained in only four sentences why there was no exception for mariculture:

> Public support for this proposal did not materialize during the comment period following the publication of the proposed regulations and the draft EIS. Cayman Farm, the only commercial mariculture enterprise concerned in this issue, provided little supportive information that they had made significant progress in achieving captive-breeding success with sea turtles. In their comment, a number of scientists expressed the view that Cayman Farm would not provide much useful information for conserving sea turtles. Finally, NMFS is concerned that sea turtle mariculture may stimulate additional commercial interest in sea turtles.[23]

The EIS found that in the three years since the first regulations were proposed, Cayman Turtle Farm had made neither significant research contributions nor sufficient progress toward its goal of self-sufficiency by 1980 to compensate for the risk that its trade would pose to wild sea turtles. NMFS noted that several scientists who had worked on the IUCN principles and recommendations submitted comments opposing an exception for mariculture: Archie Carr, George Balazs, David Ehrenfeld, Wayne King, Peter Pritchard, and Nicholas Mrosovsky. The only scientist who commented in support of mariculture was Leo Brongersma of the Netherlands Museum of Natural History.[24] Both agencies appeared to have gotten most of what they wanted. NMFS had given up the mariculture exemption it mildly supported in return for no restrictions on commercial fishing. FWS had gotten the most depleted or heavily exploited species listed as endangered and a ban on turtle mariculture.

Carleton Jones immediately filed a formal request for reconsideration because the regulations seemed to be based on a significant factual error. The farm had made its last collection of wild eggs in early 1978, and a long list of scientists and researchers had submitted letters detailing the advances they had made in such fields as endocrinology and immunology as well as animal husbandry thanks to facilities and materials provided by the farm. Was basic science not "useful information for conserving sea turtles"?[25] But Jones's request for reconsideration was denied.

The Court of Last Resort

Five days later, Jones filed a lawsuit in the federal district court in Washington, DC, seeking to overturn the regulations and to enjoin their enforcement against shipments of the farm's products. An injunction was warranted, he argued, because a mariculture ban would harm the farm irreparably. Moreover, the farm had a good chance of proving that the decision to ban mariculture was unlawful. It was contrary to the CITES policy that trade in Appendix I–listed species may be allowed if they were bred in captivity for commercial purposes. And the agencies' decision was based on factual findings that were unsupported by the information and evidence the public had submitted for their consideration. This was the very essence of an arbitrary and capricious decision.

Jones's comments must have hit their mark because the agencies quickly agreed to reconsider the listing decision and prepare a new document justifying the decision. In discussions with their lawyers, it was suggested that FWS might reconsider allowing a two-year exception for mariculture. After all, the farm had stopped taking wild eggs; the products Cayman Turtle Farm sought permission to export would be made from turtles that had been hatched and raised on the farm. Although technically speaking, these turtles were only the first filial generation, FWS's own CITES regulations allowed imports of products from turtles that were "bred in captivity." The regulations did not spell out that to qualify for an import permit, a farm would have to produce a second filial generation, animals born from parents that had themselves been conceived, hatched, and raised on the farm. Given the CITES regulations, could the agency really defend the decision?

As the FWS staff considered the agency's options, an advisor reminded them that the second conference of the CITES parties was only six months away, in March 1979. There was enough time for the United States to propose a resolution to clarify that "bred in captivity" under CITES meant animals from at least the second generation to be born on the farm. Uncertainty had crept in because the French version of the text used the phrase "elevé en captivité"—that is, raised in captivity. A clarifying resolution would have a good chance of adoption by the parties because the conference was being held in San José, Costa Rica. This meant some of the world's foremost sea turtle conservationists would be there to congratulate Costa Rica for its exemplary leadership in green turtle conservation. When it had become a party to CITES, Costa Rica had designated Tortuguero as a national park and had banned all exports of wild turtle meat and calipee from the factories in Limón. Also, the turtle conservationists would be there to lobby against Australia's proposal to move certain Indian Ocean green turtle populations from Appendix I to II so that the French turtle farm on the island of Réunion, called Farm CORAIL, could export its products made from turtles "elevé en captivité" to CITES parties, including Germany.[26]

The Australian government was still hoping to export turtle products from the farms on the Torres Strait Islands. The project had been reorganized following the parliamentary inquiry into Bob Bustard's handling of the government funds given to the applied research unit at Australian National University, which had established and ran the farming project. This controversy boiled over just after Archie Carr and Albert Main completed their report detailing some of the problems at the tiny turtle farms.[27] Farms were now receiving more technical assistance and were operating on only nine islands, each with a capacity to raise 100–500 juvenile green turtles. Although still plagued by problems of disease, parasites, and limited food supplies, the Australian government still hoped these enterprises would become financially viable and could one day be recognized as CITES-approved captive breeding facilities.[28]

If the CITES parties meeting in Costa Rica adopted a resolution strictly defining "bred in captivity" to require the reliable production of second-generation animals, this would help to reinforce FWS's interpretation and the US policy against commercial mariculture of threatened species. With luck, by the time the CITES conference was over, the DC federal district court would not yet have decided the farm's legal challenge. The government's lawyers could inform the court of the CITES decision in a supplementary legal brief, countering one of Carleton Jones's stronger arguments, and then wait for the favorable ruling.

In any event, the lawyers advised, the agencies should be sure to stress in their new decision document the significant amount of scientific information they had to support the mariculture ban. The US Supreme Court had ruled recently that when reviewing the decisions of administrative agencies such as NMFS and FWS, courts are not supposed to second-guess them, to substitute their own policy preferences for those of the agencies. When scientific facts are in dispute, courts must in fact defer to the agencies; the agencies, not the courts, have the expertise and access to the best available information. Here, the question in dispute was the effect a mariculture exemption for captive sea turtles would have on the preservation of wild sea turtles. Unless the agencies did a really poor job explaining their reasoning, the court would presume that their decision to eliminate the mariculture exemption was a valid resolution of this technical question.[29] Getting a CITES resolution strictly defining captive breeding might be icing on the cake, but it would help show the court that the decision by NMFS and FWS was rationally supported. The FWS staff decided to draft the resolution.

Finally, on December 5, 1978, three months after Cayman Turtle Farm had filed the lawsuit, the two agencies sent to the court a thirty-one-page decision memorandum signed by the directors of both NMFS and FWS. They stood behind the regulations they had issued in July: there would be no exception made for trade in green turtle products derived from mariculture. It was their opinion, based on the reams of information before them, that such trade was likely to stimulate demand for turtle products at a level that no single farm could satisfy. This renewed demand would inspire any number of new farms to get into the business by taking wild turtles and wild-laid eggs for their stock. They would then run into the same technical difficulties that Cayman Turtle Farm encountered and would never wean themselves from wild stocks, further draining the already depleted populations.[30] To prevent this from happening to sea turtles and all other endangered species, the United States would introduce a resolution at the next meeting of the CITES parties that would ensure that any captive breeding facilities that sought certification in order to trade could reliably produce enough closed-cycle product to satisfy the demand.

FWS did indeed submit the proposed CITES resolution, gambling that a supermajority of parties at the meeting would agree with this reasoning. The gamble paid off. When the United States took the proposal to the CITES conference in San José in March 1979, it passed in virtually its original form.[31] The parties agreed to Conference Resolution 2.12, calling on parties to apply a uniform interpretation of "bred in captivity" for the purposes of allowing trade in Appendix I species. In essence, the new definition said that a captive-bred animal must be the result of a captive mating by a parental breeding stock that was "capable of reliably producing second-generation offspring in a controlled environment."[32] The US delegation thought it was pretty clear that this condition was not yet the case at Cayman Turtle Farm.

Speakers in support of the resolution included Wayne King, a member of the US delegation; observers for Zambia; and Michael Bean, a lawyer for the observer, the Environmental Defense Fund, who spoke on behalf of twenty other nongovernmental conservation organizations.[33] As FWS predicted, many current and former members of the Marine Turtle Specialist Group were there and lobbied for the new definition.

One turtle specialist who did not join this effort was Joop Schulz, from the Suriname Forest Service. He and Henk Reichart, his co-worker at the Foundation for Nature Preservation in Suriname, argued that tightening the definition of "bred in captivity" posed a grave threat to sea turtle conservation programs based on captive rearing. To meet the new definition, a farm would need fifteen to twenty years before it could engage in international trade of its products, requiring a multimillion-dollar investment and no prospect of return for that length of time. Developing countries like Suriname, where many of the world's turtle populations are found, needed some promise of an economic return in order to conserve their wild resources. Prohibitory measures like hunting bans were based on wishful thinking and would not stop people from catching wild turtles.

To make the case for retaining the less restrictive definition, Schulz and Reichart described their pilot green turtle ranch. In their project, designed to capitalize on the green turtle's high reproductive potential, they took a portion of the eggs laid each season that would otherwise be destroyed by spring tides to generate a harvestable crop.[34] Every year they released a portion of the surplus yearlings and thereby restocked the wild population. But without the ability to sell turtle products abroad, they explained, there would be no money to do this.

When members of the US delegation argued that their restocking plan was unproven, Reichart pointed out that the head-started green turtles they had released had survived in the wild and joined the parental stock on the foraging grounds off Brazil, a 2,000-mile voyage from Suriname. Only time would tell whether the turtles would return to nest; it required patience and good tagging methods, both of which they had, and husbandry techniques they had learned from Cayman Turtle Farm. Wasn't the time overdue to put ranching and head-starting to the test?[35]

As a result of Schulz and Reichart's arguments, language was added to the CITES resolution making clear that the added restriction did not apply to "ranching" projects utilizing natural rookeries big enough to withstand some level of exploitation as a source of eggs instead of a

FIGURE 32 Delegates and observers at CITES Conference of the Parties on hike to Nancite Beach, Costa Rica, in March 1979 to view Pacific ridley nesting beach. From left to right: Gerald Durrell, Archie Carr, and Wayne King.
Courtesy of Peter C. H. Pritchard.

captive breeding stock. Several members instructed the CITES Secretariat to look into the possibility of developing guidelines that would allow ranching schemes for marine turtles to gain CITES approval. In this way, developing countries would have the incentive to aid the conservation of endangered sea turtles.[36] When Australia's proposal to downlist the Indian Ocean green turtle populations used by Farm CORAIL's ranching failed to pass, Schulz and Reichart realized they would have to work hard in the two years before the next CITES meeting, which would be in New Delhi, to garner support for a set of ranching guidelines that would allow their project in Suriname to succeed. Many delegates seemed to have been persuaded by the arguments of some sea turtle experts who adamantly opposed commercial use of any kind. The experts who felt differently had been too quiet, but they had begun to get an opposing view on the record. With a little luck and lots of patient persuasion they could perhaps turn the bias against captive rearing around.

As soon as the plenary session in San José was over, Michael Bean called William Butler, the Washington, DC, counsel for the Environmental Defense Fund, to tell him about the resolution. The Environmental Defense Fund was now a party to the lawsuit by Cayman Turtle Farm; the court had permitted the environmental group to intervene as a defendant. Being a party rather than an amicus—a "friend of the court" like the Cayman Islands government was—allowed Bean and Butler to make sure the agencies vigorously defended their decision to prohibit mariculture. They would now augment that defense with late-breaking news of the CITES resolution defining "bred in captivity." And the Environmental Defense Fund could make sure the court understood on which side the weight of scientific opinion rested regarding the conservation value of Cayman Turtle Farm.

With his duty to give deference to the agencies' technical judgment and the news of the CITES development, Judge John Pratt had little trouble deciding how to rule. In late May 1979, Judge Pratt issued his opinion rejecting all three of the farm's legal challenges. He found that the ESA did in fact apply to captive-bred sea turtles which are hatched and raised in a controlled environment. The comprehensive sweep of the act and the broad discretion afforded the secretaries of interior and commerce made clear that Congress had intended to include such animals even when they are no longer wild. Moreover, while it was true that CITES allowed for the trade in species threatened with extinction if the specimens were bred in captivity, the convention also preserved the right of parties like the United States to adopt stricter national measures, including a complete ban on such trade. And even if he were to assume for the sake of argument that the CITES criteria for importation were mandatory for the United States, the majority of the turtles that Cayman Turtle Farm hoped to export did not meet the definition of "bred in captivity" adopted two months ago at the second conference of the parties in San José. The evidence was clear that the majority of the turtles were hatched from eggs taken from the wild. If the secretaries were to exempt mariculture, as the plaintiffs argued for, they would run afoul of their duties to follow this new CITES policy.[37]

As soon as the Mittags learned of Judge Pratt's ruling, they retained a new lawyer, Richard Lapidus, to appeal the case. They could scarcely believe that the judge had bought the government's argument that their farm would adversely affect wild turtle populations. That argument relied on a long string of speculations: "If this happens, then that will happen." Regarding the farm's lack of self-sufficiency, they thought it was supremely unfair so late in the game for the United States to require that they sell only turtles produced from a second generation. When they bought the farm, CITES allowed the sale of animals "bred in captivity for commercial purposes."[38] They knew that several delegations, including the United Kingdom's, left the San José meeting believing that the new definition would not apply retroactively to existing mariculture facilities. And it was truly hurtful that the judge found, in essence, that the research at the farm was of no benefit to conservation. This was truly a slap in the face to the Mittags, who had purchased the farm to help achieve the goals of the ESA, to prevent the extinction of the green turtle.

Lapidus warned the Mittags these issues would be difficult to appeal, especially the one regarding the retroactivity of the recent CITES resolution. It might require hiring English lawyers to take depositions of the UK delegation. The US environmental organizations had fielded many observers at the meeting, and he was certain they would swear to the opposite. Their lawyers would argue that the United Kingdom's recollection was self-serving because the Cayman Islands is a British dependency. He believed he would probably have the best shot at appealing the issue of whether the ESA even applied to captive, domesticated versions of species that in the wild are at risk of extinction. It seemed to him that Judge Pratt had given the issue short shrift and that a different court might be willing to give the act a somewhat narrower interpretation.[39]

CHAPTER SIXTEEN

A Global Strategy

While Richard Lapidus readied his appeal of the district court's rejection of Cayman Turtle Farm's suit, the sea turtle conservationists celebrated their successes. By marshaling the available science and the provisions of the Endangered Species Act and CITES, they believed they had vanquished "conservation through commerce" and advanced preservation of sea turtles through total protection. Almost immediately, they turned their attention to the need for a global conservation strategy for sea turtles, informed by a thorough understanding of the recent advances in sea turtle biology. To make the strategy truly global, they decided to convene an international meeting in Washington, DC, in November 1979, using the conference facilities of the US State Department. The plan was to invite a select group of turtle scientists to present their research findings. Other speakers could assess available conservation methods, including nesting beach preserves, hatcheries and head-starting, and international trade bans. All the dicey utilization issues would be aired, including subsistence hunting and farming, but the focus would be on sea turtle biology, the status of sea turtle populations, and a conservation strategy.

Staff members of the Center for Environmental Education, the Defenders of Wildlife, and the US office of the World Wildlife Fund took the lead in planning the logistics and raising travel funds for the invited speakers. The planning committee decided to invite Henk Reichart from Suriname to give the case for farming and ranching. Kenneth Dodd, a newly hired herpetologist at the Fish and Wildlife Service, offered to present a paper reviewing whether mariculture benefited conservation. David Ehrenfeld volunteered to give an overview of the biological and ecological constraints on options for turtle conservation. Archie Carr wrote the first draft of the global strategy.[1] Henk Reichart and Joop Schulz were reluctant to attend, given the hostility expressed toward pro-farming members in San José at a pre-CITES meeting of the Marine Turtle Specialist Group, but Archie Carr assured them that no one would make any personal attacks on their characters or threaten their jobs.[2]

When Dave Owens, who was now teaching biology at Texas A&M University, got his invitation to the conference, he was happy to accept. He had done quite a bit of research on the reproductive physiology of sea turtles at Cayman Turtle Farm and believed this important line of inquiry was often overlooked by sea turtle conservationists. Reproductive biology, he thought, could be fairly characterized as the new frontier in sea turtle research, much as migration studies had been in the 1950s. He was somewhat concerned, however, that the Washington conference, which was pulled together rather quickly, might preempt the symposium he had planned for one month later, at the annual meeting of the American Society of Zoologists. He was glad the meeting planners had travel funds for him, but he wondered how they had managed to find so much support so quickly.

Owens had had a hard time getting funding for his symposium; his National Science Foundation grant proposal got negative reviews, and he suspected that his association with Cayman Turtle Farm was the reason.[3] It was true that his ulterior motive in planning the symposium was to get everyone who worked on sea turtles talking about the science again, but he believed it served the cause of conservation. He had learned in 1977 that the 1979 ASZ meeting would be in Tampa, not far from Gainesville. He immediately thought it would be an opportunity for Archie Carr and Carr's Florida associates to exchange views on the valuable research carried out on the farm by Jim Wood; Wood's wife, Fern; and the others who had used materials or facilities at the farm.

A previous effort to inform the Florida group had not gone well. In July 1976, when Ross Witham, from the marine research laboratory of Florida's Department of Natural Resources, had organized a meeting at the Jensen Beach campus of the Florida Institute of Technology, over 100 people had attended. It was clear that there was tremendous interest in sea turtle science and conservation, but not everyone was listening with an open mind. Owens gave a paper on his research with John Hendrickson on sex ratios and the endocrinology of green turtles. He had developed a blood test to measure testosterone levels because there was no external way to determine the sex of a sea turtle hatchling. After seeing the sex ratio of wild clutches that had been incubated at Cayman Turtle Farm, Owens had begun to suspect that with sea turtles, as with freshwater turtles, the temperature during incubation determined the sex of the hatchlings. When he suggested that it was not a good idea for conservation hatcheries to incubate turtle eggs in Styrofoam coolers, unless the goal was to produce only males, Ross Witham took umbrage at the remark, assuming Owens was criticizing his head-starting project, the centerpiece of Florida's turtle conservation program.[4]

Owens' goal was to generate a robust discussion of the latest scientific research free of the pressures of conservation politics and the endless wrangling over what was a good or a bad strategy to save turtles. When he was still a graduate student, Owens had been impressed by the quality of discussion at ASZ meetings. He thought a healthy dose of that kind of exchange was just what was needed in the sea turtle community.

When, in 1979, Owens suggested a special symposium on sea turtle reproduction at ASZ to his colleagues, they greeted the idea with enthusiasm. Nicholas Mrosovsky was now editing the *Marine Turtle Newsletter* as a means for the far-flung marine turtle specialists and other researchers to communicate about research methods and results. Mrosovsky thought a scientific meeting in Florida would help to improve the many conservation projects that were cropping up around the world. Mrosovsky was also now the co-chair of IUCN's Marine Turtle Specialist Group, having been appointed by Tony Mence following the untimely death of Tom Harrisson. Harrisson and Christine Fornari had been killed in a gruesome traffic accident in January 1976, while on a visit to Thailand, leaving the turtle group without its most skillful tactician and main point of contact with the IUCN leadership. In the first issue of the newsletter, Mrosovsky had included an obituary for Harrisson written by Archie Carr.[5]

Mrosovsky had had his own difficulties with Ross Witham. Witham had turned down Mrosovsky's application for a research permit from Florida's Department of Natural Resources to use a small number of loggerhead hatchlings in experiments to test Owens' belief that nest temperature determined the sex of sea turtle hatchlings. Although loggerheads were considerably more abundant in Florida, Mrosovsky had had to get hatchlings from James Richardson in Georgia and Sally Hopkins-Murphy in South Carolina. Ross Witham simply did not believe in temperature-dependent sex determination. He thought research that required sacrificing hatchlings to determine their sex was unnecessary and hard to justify while the state was negotiating with shrimp fishermen to reduce turtle drownings.[6]

After Owens' funding request was turned down by NSF, Peter Pritchard put him in touch with the Chelonia Institute of the Truland Foundation, which offered funds to supplement the assistance from the ASZ.[7] To get Carr's associates in Florida to attend the ASZ symposium, Mrosovsky suggested they dedicate the symposium to Archie Carr and ask him to give the keynote paper. When Carr got the invitation, he realized why he was being invited. He did not really want to be feted at this point; there

FIGURE 33 Spectators at daylight release of yearling green turtles from Florida's restocking program at Melbourne Beach, Florida, including Nicholas Mrosovsky (standing, leaning over camera, center left) and Archie Carr (kneeling, taking photograph of photographer), following the Jensen Beach sea turtle conference of July 1976. Courtesy of Peter C. H. Pritchard.

was still so much work to be done. But he accepted the invitation anyway and made preparations to attend.

Fateful Encounters

The planners of the World Conference on Sea Turtle Conservation in Washington, DC, had expected a big turnout. Even they were surprised, however, when more than 300 people, representing over forty different nations, came to the five-day meeting in late November 1979. Many excellent papers were presented, and observers were impressed by the number and diversity of sea turtle studies under way.[8] But when it came time to discuss a global strategy, the speakers acknowledged how difficult it was to think globally, given the tremendous geographic ranges of sea turtles and the many gaps in regional information about them. Under these circumstances, David Ehrenfeld advised the conference that a very conservative conservation strategy was in order, focused on protecting the nesting and aquatic habitat of wild populations, "using the simplest and least risky techniques of conservation." He cautioned against using newly acquired knowledge of turtle biology to attempt to manipulate populations, reflecting a philosophy similar to the dictum in medical training that requires doctors to, first, do no harm.[9]

Much had been learned since the first interregional turtle meeting at Jensen Beach. But this new knowledge suggested to some that those turtle populations that were greatly reduced by exploitation might not rebound, even with total protection of their nesting grounds. Under these circumstances, would conservative conservation methods suffice? Using mark-recapture studies on juvenile green turtles in the Hawaiian Islands with a modified self-crimping tag, George Balazs had shown that growth rates were much slower in the wild than in captivity. His results suggested that the age of maturity in cheloniids was more on the order of thirty to fifty years, not the five to six years previously assumed from captive studies.[10] Karen Bjorndal, Archie Carr's former graduate student, explained that her just-completed PhD research suggested that, as a consequence of their diet of algae and turtle grass, Caribbean green turtles could make only a modest investment of energy in reproduction, delaying maturation for decades. This population would probably be slow to recover from a decline in their numbers even if the population was fully protected.[11]

George Hughes, from South Africa, discussed the uncertainty that remained surrounding nesting cycles, despite the two- and three-decade duration of tagging programs based on the method pioneered at Sarawak in 1953. Tag returns were too limited for anyone to be sure whether females nested multiple times. To believe they nested more than once in their lives, one had to conclude that either the tag loss problem was severe or that the mortality of tagged females was extremely high. This too suggested that reproduction would not compensate for heavy exploitation of adult populations.[12]

The session on the status of sea turtle populations began with a historical review of the decline of sea turtles. Wayne King described how centuries of green turtle exploitation began with their discovery by Columbus at the Cayman Islands and culminated with their listing on CITES Appendix I in 1973. To illustrate the devastating effect that commercial exploitation had had, King showed only one graph. It showed the data on Tom Harrisson's green turtles that nested on the tiny islands off Sarawak, where the more than three million eggs collected yearly in the 1930s declined steadily to fewer than 10,000 by 1978 even though adult turtles were not killed on those islands.[13] King was now curator of reptiles at the Florida Natural History Museum in Gainesville, and he was also the deputy chair of the Survival Service Commission, effectively replacing Harrisson as Carr's main point of contact with the IUCN leadership. King's use of Harrisson's data was a poignant reminder to those who had known Harrisson of his brilliance, his consummate political skills, and his tireless efforts on behalf of the green turtle.

More than three dozen speakers followed King, reviewing the status of all sea turtles known to inhabit the eastern and central Pacific, the Indo-West Pacific, the waters of East and Southeast Asia, the Indian Ocean, and the Atlantic, all walking in the footsteps of Tom Harrisson.

The high point of the session came at the end of Barney Nietschmann's paper. He described the impact of commercialized turtling on the subsistence culture of the Miskito people and the Caribbean green turtle population. Three turtle factories had been built in eastern Nicaragua immediately following the collapse of the tripartite agreement in 1969 and the ejection of the Caymanian turtle boats. For the next seven years, the Miskito had captured up to 10,000 green turtles a year and sold them to the factories for processing and export as meat, calipee, and oil.

The results were both the loss of the traditional hunting methods and a severe reduction of the turtle population, which "began to erode the ecological and social heart of Miskito subsistence and culture."[14] The reprieve came only after the 1973 CITES and endangered species legislation in the United States, when Costa Rica, after becoming a party to CITES, declared a national park at Tortuguero and closed its factories at Limón. The Nicaraguan government then closed its turtle companies, but allowed the Miskito to resume subsistence turtling.

What Nietschmann did not tell the audience was that he was the man who had convinced Nicaragua's President Samoza to close the factories. Nietschmann had gone to Managua to visit Jaime Incer, a widely respected volcanologist from the Department of National Geography in Managua, who was seeking the president's support for designating the region around the Masaya volcano as a nature reserve and national park.[15] When Incer met with Samoza, Nietschmann went also so that he could show the president *The Turtle People*, a film made by his former student, the anthropologist Brian Weiss, about the social and ecological change that commercial turtling had brought to the Miskito of the eastern coast of Nicaragua. The Miskito villagers had stopped eating turtle meat, and the men no longer tended their vegetable gardens. Their consequent dependence upon purchased food made them feel impoverished in their new cash economy.[16] It was as if the nation had declared war on the Miskito, destroying their subsistence culture by building the factories at Bluefields and Puerto Cabeza.[17]

Now, said Nietschmann, with CITES and the ESA, another pattern of change was coming to the turtle people, a change guided by the desire to prevent the green turtle's extinction. He had once asked who would kill the last turtle. His question now was what would happen to those people whose cultures are adapted to these animals. What would the Miskito lose if sea turtles were lost to them through prohibitive legislation? One thing was true for both the endangered culture of the Miskito and the endangered species of sea turtles: neither could be saved through commerce. Nietschmann said, "Just as Miskito society was not saved by selling a subsistence resource, endangered and threatened species cannot be saved by selling them."[18]

The conference session on conservation theory, techniques, and law began with David Ehrenfeld's review of the limited tools available for conserving sea turtles. The limitations were due to the "mysterious life cycles and obscure ecological relationships, long migrations across international boundaries, unknown population dynamics, unknown taxonomic relationships of different populations, nesting cycles of highly variable length, and an exceedingly long maturation time."[19] In view of these unknowns and the constraining aspects of sea turtle biology, he concluded that the best option was to do everything to protect the remaining wild populations, eschewing risky manipulations like long-range transplantation, head-starting, artificial incubation at different temperatures to adjust sex ratios, captive breeding to maintain gene pools, and management for maximum sustainable yield. Commercial farming and ranching were also beyond the pale, given their likelihood of causing a net drain on wild populations. Erhenfeld said he would not reprise his previous arguments about mariculture but referred the audience to his past papers.[20] But he had a particular message for the proponents of cottage-industry ranching or head-starting. First, they should heed Nietschmann's lesson about the introduction of cash payments for a resource into a subsistence culture. Ehrenfeld concluded: "There is a danger in teaching people that conservation is always accompanied by a cash profit, and . . . it is very wrong also to assume the superior attitude that peoples in poor countries are incapable of having or acquiring moral feelings of conservation."[21]

As Henk Reichart got up to speak, he sensed that Ehrenfeld's last comment had been aimed at him. Ever since the San José CITES meeting in March, he had been arguing for exactly what Ehrenfeld now characterized as a moral danger. Reichart believed passionately, as his mentor Joop Schulz had taught him, that international policy and a world strategy for conservation of sea turtles must allow developing countries to realize some economic benefit from their wild resources. They could not afford to engage in wildlife conservation just for the sake of it; they had to make an economic return. Where, as in Suriname, the local sea turtle population was robust enough or had lots of doomed nests, ranching to support hatcheries and head-starting could take advantage of the sea turtle's large reproductive potential and allow local people to benefit, instead of forcing a hands-off policy on them.[22]

When Nicholas Mrosovsky heard Reichart's paper, he was reminded of a conversation with Joop Schulz he had late one night on a nesting beach in Suriname. Schulz was describing the program he had devised to manage the exploitation of green turtles. Beginning in 1965, his forest service had declared the main nesting beaches as nature reserves, substituting nest monitoring, tagging, and beach patrols for egg and adult-turtle harvesting. But egg collecting had later resumed with the approval of the forest service, in part to generate revenues to pay for the conservation programs. When Schulz asked Mrosovsky how many eggs his conservation program could safely take, Mrosovsky had demurred. He was not sure it was safe to take any number of eggs; at least, that was what he remembered Carr had said at Tortuguero. But Schulz had told him in no uncertain terms that he was not asking him *if* they should take eggs, just how many. The realities of the Suriname social and economic conditions had just as much bearing on the conservation strategy as did the biology of the sea turtles.[23]

That conversation on Bigasanti beach ten years before was the beginning of Mrosovsky's realization that he no longer believed that total protection was the best course for marine turtles. He knew that other turtle scientists who felt the same way, especially John Hendrickson, had been ostracized by Carr and his associates; and he wondered how far into the wilderness he would be banished now that he was beginning to air his contrarian views in editorials in the *Marine Turtle Newsletter*.[24] The preceding year, Wayne King, in his new role as deputy chair of the Survival Service Commission, had reversed Mrosovsky's appointment as co-chair of the Marine Turtle Specialist Group, replacing him with George Balazs from Hawaii, who had campaigned almost as vigorously against Cayman Turtle Farm as King had.[25] In the latest issue of the newsletter, issue 13, which was distributed at the conference, Mrosovsky had penned an editorial urging the World Conference delegates to debate the relative risks and benefits, as conservation strategies, of managing the controlled use of turtles as resources versus translocation, artificial hatcheries, and head-starting.

It did not appear, however, that either his editorial or Reichart's paper on Suriname had opened anyone's mind to the idea of managed commercial utilization of turtles, even in developing countries. But if they had, the paper that followed, by Kenneth Dodd, the FWS herpetologist, would have promptly closed those minds again. Dodd pointed out that turtle farms and ranches seemed to be starting up all over the world and would be seeking approval as captive breeding facilities under CITES. This unfortunate development vindicated the recent policy findings by the US government that mariculture does not benefit conservation but instead stimulates markets and the proliferation of farms and ranches. He went on to say that "sea turtle farms and ranches are based on false premises about the extent of biological knowledge of the species" and about the applicability of the concept of optimum yield to sea turtles, citing a recent paper by Joop Schulz.[26] As to the value of the research on farms, Dodd said that of the sixty-one research projects cited by Cayman Turtle Farm in its lawsuit challenging the US import ban, only four related directly to conservation.[27]

Just before the discussion of conservation strategy, Nicole Duplaix and David Mack, two staff members of the US office of TRAFFIC, the wildlife trade monitoring group, presented a review of world trade data. Wayne King had suggested that they analyze the available trade statistics to show which countries were importing and exporting sea turtle products. This would allow the conference to assess the impact of CITES on international trade, the major driver of sea turtle exploitation, and to see whether the parties to the convention were complying with their duties to compile and report these data. Duplaix and Mack showed that there was still a remarkable amount of international trade in sea turtles despite the inclusion of almost all sea turtles on Appendix I of CITES. This was because the major international trading nations were not yet parties to CITES or had taken reservations for sea turtles, exempting themselves from the trade restrictions. It was clear that many parties were not adequately implementing the treaty's requirements. This allowed a sizable trade to continue, especially the leather trade, which was draining populations of Pacific ridleys and green turtles, and the tortoiseshell trade, the main reason hawksbill turtles were in trouble.[28]

Before the conference, the planning committee agreed reluctantly with Peter Pritchard's suggestion that representatives of the turtle industry be invited to speak at the conference and take questions. Besides presenting a paper on their research on captive reproduction and nutrition, Jim and Fern Wood were asked to lead a roundtable discussion on farming. The owners of Cayman Turtle Farm could participate in this roundtable although they would not be on the formal program. Judith Mittag was glad she had accepted this half-hearted effort at inclusion when she heard

Duplaix and Mack's paper. Seemingly out of the blue, they suggested that Cayman Turtle Farm was contributing to the illegal trade in turtle products by buying wild-caught turtle meat from the leather factories in Ecuador, labeling it "captive-bred and reared" and selling it in Europe.[29]

When it was Mittag's turn to speak at the roundtable, she explained that the farm's principal goal was to conserve green turtles and that it would do this by marketing farm-reared products to supplant uncontrolled and destructive harvests of wild turtles.[30] She emphasized that the farm had achieved self-sufficiency as required by the captive-bred provision of CITES; the last collection of eggs from Suriname was in 1978. She realized that there was disagreement about the meaning of the term "captive-bred," but fortunately, the CITES management authority of the United Kingdom agreed that the farm's products met the definition. The farm would continue and likely expand its trade with European soup manufacturers and others, now that the turtle processing factories in Nicaragua and Costa Rica were shut down. Johnny Johnson, who was still the farm's chief operating officer, said that with the closing of the US market under the ESA regulations, the farm would have to increase its sales and expand its product line in order to cover expenses.

During the question and answer period, Mittag was asked about the allegations in Duplaix and Mack's paper that the farm was laundering turtle meat from Latin America. Now Mittag was furious. She replied with control, however, stating calmly that this was an outright falsehood designed to discredit the farm, like so much else that had been said that day. Why would the farm need to import turtle meat when they had so much stock? The illogic of the allegation was further evidence of its fabrication. She had wanted to say that the American environmental organizations were using the US regulations to treat the farm as if they were heroin dealers.

Johnson was convinced that the farm was the focus of so much irrational hatred because it was such an easy target. A foreign-owned company making ingredients for gourmet soup and luxury products, it was easier to vilify than the thousands of commercial fishermen who were drowning sea turtles in their nets to satisfy the American appetite for shrimp. But it would not do to say this out loud. He did say, however, that if the allegations of meat laundering were not retracted, the farm would sue the conference organizers for libel.[31]

At the coffee break, Mittag was surprised to see Archie Carr walking in her direction with the obvious intent of speaking to her.[32] Carr wanted to reassure her that he did not believe the farm was laundering turtle meat. He was ill at ease, however, and for once in his life, his way with words seemed to fail him. Carr said he knew her husband Heinz was a prominent international lawyer and would not be so stupid as to do something like that. Judith Mittag was so outraged she could not answer. It seemed her role in the farm was invisible to this man of whom she had wanted to ask so many things. Was she mistaken or had he not recommended farming back in 1967 in his book *So Excellent a Fishe*? She had read and thoroughly enjoyed the book when she and Heinz were deciding whether to buy the farm. And how could he say that their program of releasing yearling green turtles in the waters around Grand Cayman and off Suriname was not helping to conserve wild turtles? How was their program different from the US–Mexico experimental head-starting project for the endangered Kemp's ridley turtle, which Carr supported? There were so many things she wanted to ask the foremost expert on sea turtles. But his comment had left her speechless, and the moment was lost.

Before the end of the meeting, Mrosovsky learned just how far into the wilderness he would be sent. For several months he had been negotiating with the executive secretary of the Survival Service Commission to increase its financial support of the *Marine Turtle Newsletter*. Tom Harrisson had been enthusiastic about the newsletter and had promised to get money for it from IUCN. His untimely death had brought an end to those efforts, and Mrosovsky had had to scrape together support from the World Wildlife Fund–Canada and from his own department at the University of Toronto. But he was not quite as diplomatic as Harrisson and had balked at the suggestion of the Survival Service Commission that the newsletter represent only the consensus view of the Marine Turtle Specialist Group with respect to issues of conservation. He believed firmly that a diversity of views should be aired in the newsletter, especially on the topic of conservation through utilization.[33]

When a group of conference delegates recommended establishing an editorial committee for the newsletter to ensure that it reflected IUCN policies, Mrosovsky dug in his heels. IUCN could keep its paltry sums; the newsletter would remain independent. He thought he had prevailed when, despite his rejection of the editorial committee, the delegates nevertheless adopted in the conservation strategy an action item calling on the "IUCN/SSC Marine Turtle Newsletter [to] make biologists and government conservation officials aware of the latest information on sea turtle conservation, management, and research and the status of implementation of the Sea Turtle Conservation Strategy." But when the proceedings of the conference were published, he was quite surprised to find that the Sea Turtle Conservation Strategy made no reference to the newsletter. The action item now read: "Governmental and nongovernmental conservation agencies and organizations should make biologists and governmental conservation officials aware of the latest information on sea turtle conservation, management, and research, and the status of implementation of the Sea Turtle Conservation Strategy through newsletters and other media."[34]

When Mrosovsky learned that his newsletter had been written out of the strategy, he decided to accept Cayman Turtle Farm's offer of support. Judith Mittag's offer came with no strings attached, and he much preferred that to being beholden to the conservation orthodoxy.

"A Relatively Small Taxon"

One month after the World Conference, it was time for the ASZ meeting in Tampa, and Dave Owens was wondering if his special symposium had been such a good idea after all. The conference in Washington, DC, had degenerated at times into name-calling and allegations of illegal behavior. Everyone who was ever associated with Cayman Turtle Farm was seen as an enemy of the green turtle, including his mentor John Hendrickson and his friends Glenn Ulrich and Jim and Fern Wood, with whom he had worked to make the breeding program a success. Mrosovsky was only the latest angel to be cast out from Heaven.[35]

But it turned out to be a very nice symposium, and in the end Owens was grateful that Mrosovsky had suggested they make it a celebration of Carr's career. Carr gave a wonderful paper on the many remaining puzzles of what he called the "ecologic geography" of sea turtles. Carr's former students and Florida colleagues Peter Pritchard, Harold Hirth, and Ross Witham gave papers that revealed how much he had inspired them to do the very best work they could. John Hendrickson gave one of his classic "big picture" papers designed to provoke and inspire more work that would at last explain the intriguing evolution of the sea turtles.[36]

Nicholas Mrosovsky read aloud the dedication to Archie Carr that he had offered to prepare. On the occasion of his seventieth birthday, it was fitting to honor Carr for giving birth to the idea that sea turtles could and should be saved and for his outstanding contributions to conservation and zoology. While many questions of sea turtle biology remained to be answered, Carr's classic study of the green turtle colony at Tortuguero had formulated all the questions, and the sea turtle community was well on its way to answering them.[37] In his own, separate conference paper, Mrosovsky acknowledged that deep divisions sometimes arose over priorities and the best means of conserving turtles, and that there was a risk in following Carr's paradigm indefinitely.[38] But if there was ever any ill will, it was not because of Carr. He had "participated in these debates, vigorously at times but with exemplary lack of malice."[39]

Although embarrassed by the dedication, Carr enjoyed being in the company of others who saw the importance of research on such a "relatively small taxon." He joined the discussion of the work Cayman Turtle Farm had facilitated as if it was indeed the best available science.[40] There was none of the rancor that had been in the air in Washington, DC; it was time to celebrate science and the sea turtle, and to be together once more in the brotherhood of the green turtle.

"The Logic of This Case Is Appalling"

A year later, in 1980, the court of appeals handed down its decision in Cayman Turtle Farm's suit challenging the US ban on import of cultured turtle products. The court agreed with Judge Pratt's analysis that under the plain language of the ESA, the secretaries of interior and commerce clearly had the authority to regulate domesticated animals of species listed as endangered or threatened.[41]

The farm's appeal had failed to convince the court, but not entirely. In a separate, unpublished opinion, Judge George Mackinnon concurred with the result but wrote:

> Given the fact that a very substantial demand for products of green sea turtles grew up based almost exclusively on such turtles captured in the

wild and killed, the conclusion that a commercial farming operation enhances the propagation and survival of the species in the wild by lessening the demand upon wild stocks has, in my opinion, more support for it than the contrary conclusion that the sale of products from such an operation increases the appetite and demand for such products so greatly that the threat or danger to the species in the wild is increased. My disagreement with the Secretaries of Interior and Commerce over the effect of granting a mariculture exemption goes to the substantiality of the evidence in the record to support their decision not to include such an exemption in their final regulation, and to whether that decision was arbitrary and capricious. Although appellant Cayman Turtle Farm, Ltd. argued in the district court that the administrative record did not support the Secretaries' decision, on this appeal it has argued only that the Endangered Species Act of 1973 does not apply to farmed animals. Since issues not raised on appeal are abandoned, . . . I agree that the district court's judgment should be affirmed.

Attorney Lapidus had chosen the wrong issue to appeal. He had not challenged the agencies' conclusion that, based upon the best scientific information available, the farm would do more harm than good to sea turtle conservation. Had he done so, the farm would have won.

When Lapidus sent the opinion to the Mittags, they felt defeated and sad by the time they finished reading it. They could no longer afford to carry the farm's expenses. Now, they had no choice but to reduce the size of the breeding herd by slaughtering the older ones and releasing the younger ones. They would sell the farm to the Cayman Islands government, to be run as a tourist attraction only. Gone were their dreams of celebrating the beauty and magnificence of the green turtle while helping to sustain its survival as a species.

They simply could not fathom the reasoning of the Americans and their laws that were supposed to save species from extinction. The Mittags might have felt a bit of comfort had they been able to read Judge Mackinnon's personal notes on the case. In the cover note to his fellow judges when circulating his draft opinion, he simply said: "The logic of this case is appalling. The agency contends that breeding large numbers of endangered species *in captivity* and thus greatly increasing the number of such species tends to further endanger the species."[42]

The transformation of the green turtle from food to icon was thus affirmed by the narrowest of margins. There would be other acts and players drawn into the drama. But for the connoisseurs of green turtle soup, and the proponents of conservation through commerce, the play was over.

Supply and Demand

The Mittags sold the turtle farm to the Cayman Islands government in 1983. The farm continued to sell turtle meat to local restaurants and expanded the Mittags' head-starting program, releasing surplus hatchlings and yearlings to restore green turtles to the Caribbean and as a popular tourist activity. The Cayman Islands and UK governments sought approval to export green turtle products under CITES several times in the ensuing years but never gained sufficient support. Prospects for approval appeared best after a delegation of congressmen from the United States toured the farm and agreed to hold a hearing on the issue of turtle farming and ranching. The Fish and Wildlife Service proposed a change to the special rules to allow exports through the United States. The regulations were withdrawn, however, after the UK effort to register the farm as a captive breeding facility for long-maturing species failed at the CITES conference in Gaborone, Botswana, in 1983.

In subsequent years, employees of the Cayman Turtle Farm observed green turtles mating in waters offshore of the farm and females nesting on Grand Cayman's Seven Mile Beach. All were identified by a living tag devised by John and Lupe Hendrickson and applied to the yearlings prior to their release. These observations were reported in the *Marine Turtle Newsletter*, which continues to be published by volunteer editors from the sea turtle research community and celebrated its 100th edition in 2002. Waves generated by Hurricane Michelle in November 2001 inundated the breeding lagoon and rearing tanks at the farm, releasing thousands of green turtles, including three-quarters of the breeding herd, to the sea. The facility was rebuilt further inland, expanded, and reconfigured as a nature park, offering visitors an opportunity to swim with the turtles in an artificial reef lagoon.

The small turtle farms on nine of the Torres Strait islands continued for a few more years, but government funding ended in 1980 when it grew increasingly difficult to find fish for the hungry turtles and to keep them free of disease and parasites. Barney and Judith Nietschmann visited the Torres Strait Islands in 1976 to continue their ethnographic and ecological studies of traditional maritime cultures. Barney studied the hunting and ecology of dugongs and green turtles and their role in the Torres Strait society and economy. When the Nietschmanns toured several islands, they saw signs of the former farm project, but it was not a topic of general conversation.

In 2004, the IUCN Marine Turtle Specialist Group, whose members now numbered in hundreds, classified green turtles for the Red Data Book as endangered globally. The recommendation was based on evidence of a 48 to 65 percent decline in the number of mature females over the past century and a half at thirty-two index nesting sites, as required by the IUCN's new quantitative criteria for classifications. The index sites included the nesting beaches at Tortuguero, Raine Island on the Great Barrier Reef, and Telang Telang Besar, on Sarawak's Turtle Islands. Nesting colonies on Sarawak's islands are protected by a Wild Life Protection Ordinance prohibiting the taking of all marine turtles and their eggs. The Sarawak Museum's Turtle Board coordinates research and operates conservation hatcheries on the islands, which are included in the Talang-Satang National Park. In a project Tom Harrisson would have enjoyed, the Turtle Board has placed approximately 500 reef balls on the seabed around the islands to deter illegal trawling, enhance tourism diving, and provide shelter for resting turtles. The Sarawak Forest Service's marine department is establishing farms to produce soft-shelled terrapins for the export market.

Because all the turtle farms and ranches either closed or changed their function, the question was never resolved whether turtle mariculture would trigger an increase in poaching of wild turtle populations or would flood the market and thus reduce pressure on wild populations. The issue was moot once the taste for turtle soup and the demand for turtle products declined. Coastal tourism became a global industry and recreational diving and snorkeling brought people in contact with marine turtles in the wild. In 1986, Clem Tisdell at the University of Queensland published the only economic analysis ever done on the question; he concluded that the impact of farming on the level of wild harvesting could be "small or of zero magnitude."[1] Its effect depended on the elasticity of supply of turtles in the wild. In other words, it was all a matter of supply and demand.

Appendix A

The 1966 U.S. Classification of *Chelonia mydas* as Rare and Endangered

From US Department of Interior, Fish and Wildlife Service, *Rare and Endangered Fish and Wildlife of the United States* (the "Red Book"), Resource Publication no. 34 (1966).

Peripheral*

GREEN TURTLE   *Chelonia mydas mydas* Linnaeus
Order   CHELONIA   Family   CHELONIDAE

Distinguishing characteristics: Abutting, non-overlapping shields on carapace; four pairs of lateral shields in shell; one pair of prefrontals on head; very large size; paddlelike feet.

Present distribution: Tropical oceans in shore waters. Wanders up United States coast during summer.

Former distribution: Same; but uses Florida beaches as a laying site.

Status: "Practically extirpated as a breeding entity in the fauna of the U.S." (Carr and Ingle). Still common as a breeding entity on Hawaiian Islands (RLW).

Estimated numbers: In U.S., probably very few, but world-wide, still fairly abundant.

Breeding rate in the wild: Probably once a year, although possibly once every two years, 125–200 eggs per female.

Reasons for decline: Widely used for food; young subject to very heavy predator and human pressure.

Protective measures already taken: Molestation of nesting sea turtles and their eggs is prohibited in South Carolina, Georgia, Florida, Texas, Puerto Rico, and Hawaii. Hatchlings are flown from Caribbean beaches to Florida for release. Similar release techniques are employed in Buck Island Reef National Monument, Virgin Islands, and in Virgin Islands National Park. Results in the form of return of released hatchlings not verified as yet.

Measures proposed: Raising hatchlings in impoundments up to shell lengths of 6–8 inches, then releasing them (Carr and Ingle); establishment of protected breeding beaches.

Numbers in captivity: Very many, practically every saltwater aquarium and a few zoos have them often as many as 5–10.

Breeding potentiality in captivity: Practically nil; no opportunity to lay eggs.

References:

A. Carr and R. Ingle; The Green Turtle in Florida; Bull. Marine Sci. Gulf and Carib., 1959, pp. 315–20.

Ronald L. Walder (pers. comm.).

* The classification "peripheral" was defined as a species "whose occurrence in the United States is at the edge of its natural range and which is rare or endangered within the United States although not in its range as a whole."

265

Appendix B

IUCN Principles and Recommendations on Commercial Exploitation of Sea Turtles

As reprinted in the first issue of the IUCN/SSC *Marine Turtle Newsletter*, no. 1, August 1976 (editor: N. Mrosovsky, University of Toronto) from the *IUCN Bulletin 6*, no. 4 (April 1975).

1. Because the majority of the distinct populations of Chelonia (green turtles) are extinct, threatened or rapidly declining, the entire group should be considered endangered.
2. The reasons for the extinction and decline of populations include particularly exploitation for meat, hides, eggs and other products (including souvenirs), massive killing of turtles in the trawl nets of fishing fleets as well as increasing habitat destruction and disturbance.
3. The situation has become even more critical with the expansion of international commercial trade in sea turtles and their products.
4. As regards trawling, urgent attention should be given to encourage the use of nets designed to minimize undesirable catches of turtles, and research into this question should be given funding priority.
5. As regards souvenirs, the taking and preparing of turtles and turtle products for the primary purpose of souvenirs should be strongly discouraged.
6. As regards primary exploitation (meat, hides, eggs), where it can be demonstrated that local turtle populations can tolerate exploitation, and the desire or necessity is present, this should be done only by peoples traditionally dependent on them, with methods ensuring minimal waste and for local utilization. The diversion of wild sea turtle resources from traditional use by local people, or the expansion of that use, to satisfy or extend the demands of international commerce, is condemned.
7. It is emphasized at this point that there is a distinction between turtle farming and turtle ranching; a turtle farm implies that the unit is completely independent of wild stocks; a turtle ranch is a unit dependent on wild populations for eggs or turtles with the animals kept in varying degrees of captivity (H. Hirth, FAO Fisheries Synopsis No. 85, "Synopsis of Biological Data on the Green Turtle," December 1971).
8. Further, in recognition of the deteriorating energy and food resources of the world, it is advocated that wherever possible any turtle culture be maintained at the lowest applicable trophic level.*

* All organisms are classified as producers, primary consumers (herbivores), secondary consumers (carnivores), or decomposers according to the place they occupy in the food chain of an ecosystem. The placement is termed "trophic level." Therefore, herbivorous species should subsist on a diet based on plant protein and carnivorous species on animal protein.

9. Farming objectives which lead to the expansion of existing markets resulting possibly in an increased exploitation of wild turtles are unacceptable. However, it would be consistent with the foregoing principles to accept turtle farming whose products will replace wild turtle products in existing traditional markets. The acceptability of any farm should be demonstrated by suitably designed and independently evaluated tests and data. Moreover, those ranching endeavors satisfying the above conditions and which can be shown not to harm wild turtle populations are also acceptable.

10. Funds should be provided for the preparation of informative pamphlets to promote the application of the foregoing principles and immediate measures should be taken to ensure the early implementation of such action as is necessary to conserve the marine turtle resource in accordance with these principles.

11. Nearly all the considerations stated for Chelonia may be applied with equal force to populations of the six other species of marine turtles.

Notes

Introduction: From Seafood to Icon

1. Nicholas Mrosovsky noted that sea turtles had displaced whales in the environmental conservation effort, in his book *Conserving Sea Turtles* (London: British Herpetological Society, 1983), 1. The sea turtle is also a symbol for resistance to economic globalization and a source of instant "green" credibility for election campaigns. In the 2007 film *Battle in Seattle*, depicting historical globalization protests at the meeting of the World Trade Organization in 1999, some protesters wore sea turtle costumes similar to ones used to protest proposals to delist hawksbill turtles at meetings of the CITES parties.

2. The abbreviation CITES stands for Convention on International Trade in Endangered Species of Wild Flora and Fauna.

3. See, e.g., Annette C. Broderick et al., "Are Green Turtles Globally Endangered?" *Global Ecology and Biogeography* 15 (2006): 21.

4. Steven L. Yaffee, *Prohibitive Policy: Implementing the Federal Endangered Species Act* (Cambridge: MIT Press, 1982), 164.

5. 16 U.S.C. sec. 1533. See Holly Doremus, "The Purpose, Effects, and Future of the Endangered Species Act's Best Available Science Mandate," *Environmental Law* 34 (2004): 397–450.

6. See, e.g., Graham Webb, "Conservation and Sustainable Use: An Evolving Concept," *Pacific Conservation Biology* 8 (2002): 12–26; Jon M. Hutton and Nigel Leader-Williams, "Sustainable Use and Incentive-Driven Conservation: Realigning Human and Conservation Interests," *Oryx* 37, no. 2 (2003): 215–26.

7. See, e.g., John R. Hendrickson, "Marine Turtle Culture: An Overview," *Journal of the World Aquaculture Society* 5, no. 1-4 (1974): 167–81. Cf. David W. Ehrenfeld, "Conserving the Edible Sea Turtle: Can Mariculture Help?" *American Scientist* 62 (1974): 23–31.

8. Frederick R. Davis, *The Man Who Saved Sea Turtles: Archie Carr and the Origins of Conservation Biology* (New York: Oxford Univ. Press, 2007).

9. Lisa M. Campbell, "Science and Sustainable Use: Views of Marine Turtle Conservation Experts," *Ecological Applications* 12, no. 4 (2002): 1229–46. The story of the Cayman Turtle Farm's tireless efforts to gain acceptance from Carr

and turtle conservationists is recounted in careful detail in Peggy Fosdick and Sam Fosdick, *Last Chance Lost? Can and Should Farming Save the Green Sea Turtle? The Story of Mariculture, Ltd.—Cayman Turtle Farm* (York, PA: Irvin S. Naylor, 1994).

10. See Florida Fish and Wildlife Research Institute, "Green Turtle Nesting in Florida," n.d., Florida Fish and Wildlife Conservation Commission website.

11. This colorful fishery was captured in popular periodicals ranging from *National Geographic* to the *New Yorker* and inspired an award-winning work of fiction, *Far Tortuga*, by Peter Matthiessen. See Peter Matthiessen, "To the Miskito Bank," *New Yorker*, Oct. 27, 1967.

12. See Barnard Q. Nietschmann, "When the Turtle Collapses, the World Ends," *Natural History* 83, no. 6 (1974): 34–43.

13. Robert E. Schroeder, "Buffalo of the Sea," *Sea Frontiers* (May/June 1966): 176–83.

14. Archie Carr, *The Reptiles* (New York: Time-Life Books, 1963), 175; and Carr, *The Windward Road: Adventures of a Naturalist on Remote Caribbean Shores* (New York: Knopf, 1956).

15. Carr, *The Reptiles*, 175.

16. Ibid., 176.

17. Private papers of George MacKinnon, appellate judge on the court of appeals for the DC Circuit, Minnesota Historical Society Library.

18. See Judith M. Heimann, *The Most Offending Soul Alive: Tom Harrisson and His Remarkable Life* (Honolulu: Univ. of Hawaii Press, 1998).

19. Tom Harrisson's wife, Barbara Harrisson, had an even greater impact on the field of species survival, particularly primates. See Barbara Harrisson, *Orang-utan* (London: Collins, 1962). Concerned over the plight of the orangutan of Borneo, she convinced the American Association of Zoological Parks and Aquariums to adopt an agreement not to buy, accept, or trade in any orangutans, rhinoceroses, giant tortoises, or other rare species placed on a list the association kept and added to by a two-thirds majority vote. This agreement was the precursor to CITES, the Convention on International Trade in Endangered Species, adopted in Washington, DC, in 1973. David W. Ehrenfeld, *Biological Conservation* (New York: Holt, Rinehart & Winston, 1970), 109.

Chapter 1: Turtle Kraals and Canneries

1. Charles H. Stevenson, "The United States Fish Commission," *North American Review* 176, no. 557 (1903): 594. George Brown Goode—who was the assistant to the first commissioner, Spencer F. Baird, and was later himself the

commissioner—had the idea of working with the tenth census in 1880 to prepare an exhaustive survey of the fish and fisheries resources of the entire US coast. The resulting work was G. Brown Goode, ed., *The Fisheries and Fishery Industries of the United States,* 7 vols. (Washington, DC: Government Printing Office, 1884–87).

2. Quoted in Robin W. Doughty, "Sea Turtles in Texas: A Forgotten Commerce," *Southwestern Historical Quarterly* 88 (1984): 43.

3. Ibid., 54.

4. Charles H. Stevenson, "Preservation of Fishery Products for Food," *Bulletin of the U.S. Fish Commission* 18 (1899): 539.

5. Henry A. Hildebrand, "A Historical Review of Sea Turtle Populations in the Western Gulf of Mexico," in *Biology and Conservation of Sea Turtles,* ed. Karen A. Bjorndal, rev. ed. (Washington, DC: Smithsonian Institution Press, 1995), 447–53.

6. Ibid., 451.

7. Doughty, "Sea Turtles in Texas," 49–50, 54. Doughty cites Hildebrand, "Historical Review," 451, as the source regarding sea turtles feeding on packing-house wastes.

8. Doughty, "Sea Turtles in Texas," 55.

9. Ibid., 50.

10. Hildebrand, "Historical Review," 451.

11. Wayne N. Witzell, "The Origin, Evolution, and Demise of the US Sea Turtle Fisheries," *Marine Fisheries Review* 56, no. 4 (1994): 22.

12. Stevenson, "United States Fish Commission," 593–601.

13. Doughty, "Sea Turtles in Texas," 64. Doughty notes that Isaac Kibbe was the first state official to call for conservation and management measures for the green turtle.

14. John J. Brice, "The Fish and Fisheries of the Coastal Waters of Florida," in US Commission of Fish and Fisheries, *Report of the US Commissioner for the Year Ending June 30, 1896,* App. 6 (Washington, DC: Government Printing Office, 1897), 341.

15. W. L. Hobart, ed., *Baird's Legacy: The History and Accomplishments of NOAA's National Marine Fisheries Service, 1871–1996,* NOAA Technical Memorandum NMFS-F/SPO-18 (Washington, DC: US Department of Commerce, June 1996), 5.

16. The study was the seven-volume *Fisheries and Fishery Industries of the United States;* see note 1, above.

17. "The New Fish Commissioner: President Cleveland Ends the Suspense by Appointing Commissioner Brice to the Vacant Office," *New York Times,* March 17, 1896, 4.

18. James J. Parsons, *The Green Turtle and Man* (Gainesville: Univ. Press of Florida, 1962), 25.

19. Archie Carr, *Handbook of Turtles: The Turtles of the United States, Canada, and Baja California* (Ithaca, NY: Cornell University Press, 1952), 355. In the early 1670s, the lord proprietors of the Carolinas asked their counterparts in the Bahamas to submit a bill to the English Parliament "for the preservation of turtle" and then repeated the request five years later, when it went unheeded. Parsons, *Green Turtle and Man*, 24.

20. Parsons, *Green Turtle and Man*, 25.

21. Carr, *Handbook of Turtles*, 353.

22. Archie Carr and David K. Caldwell, "The Ecology and Migrations of Sea Turtles, 1: Results of Field Work in Florida, 1955," *American Museum Novitates*, no. 1793 (1956): 6–7.

23. Frank Lund, "Florida's Turtle Fisheries" (term paper, Dept. of Wildlife Ecology and Conservation, University of Florida, 1978), citing C. E. Carter and J. P. Bloom, eds., *Territorial Papers of the United States*, vol. 22, *Territory of Florida* (Washington, DC: Government Printing Office, 1934–75).

24. Ibid.

25. John J. Brice, "Fish and Fisheries of the Coastal Waters of Florida," 312.

26. J. W. Collins, "Report on the Discovery and Investigation of Fishing Grounds, Made by the Fish Commission Steamer *Albatross* during a Cruise along the Atlantic Coast and in the Gulf of Mexico: with Notes on the Gulf Fisheries," in *Report of US Commission on Fisheries for the Year 1885*, app. 14 (Washington, DC: Government Printing Office, 1887).

27. Ibid.

28. Brice, "Fish and Fisheries of the Coastal Waters of Florida," 313.

29. W. A. Wilcox, "Commercial Fisheries of Indian River, Florida," in *Report to the US Commissioner of Fish and Fisheries for the Year Ending June 30, 1896*, app. 5 (Washington, DC: Government Printing Office, 1898), 253.

30. Parsons, *Green Turtle and Man*, 26.

31. Osha Gray Davidson, *Fire in the Turtle House: The Green Sea Turtle and the Fate of the Ocean* (Cambridge, MA: Perseus Books, 2001), 70–71.

32. Joyann Bodden, "Captain Allie's Memories: Looking Back on 55 Years at Sea," *Northwester*, May 1973, 51–52. Captain Allie Ebanks recalled using compensation money earned when the US government commandeered his turtle schooner during World War II to buy the *A. M. Adams*; he named the vessel after his good friend, the manager of the cannery in Key West. A certificate of registry located in the Cayman Islands National Archives indicates that Mrs. Irene Elizabeth Ebanks was the owner and Thompson Fish Company of Key West was the mortgage holder.

33. William C. Schroeder, "Fisheries of Key West and the Clam Industry of Southern Florida," in *Report of the US Commissioner of Fisheries for 1923*, appendix 12 (1924), 50–53.

34. Ibid.

35. Ibid., 51.

36. Ibid.

37. Brice, "Fish and Fisheries of the Coastal Waters of Florida," 287, 297.

38. Brice's source of recommendations for Florida's turtle fisheries, Ralph Munroe, was experimenting with growing sheepshead sponges and had demonstrated its feasibility.

39. Brice, "Fish and Fisheries of the Coastal Waters of Florida," 312–13.

40. Ibid., 341.

41. Witzell, "US Sea Turtle Fisheries," 22.

42. Ralph M. Munroe, "The Green Turtle, and the Possibilities of Its Protection and Consequent Increase on the Florida Coast," *Bulletin of the U.S. Fish Commission* 17 (1897): 274–74.

43. Carr, *Handbook of Turtles*, 171.

44. Ibid.

45. Ibid., 169.

Chapter 2: Turning Turtles on the Great Barrier Reef

1. Parsons, *Green Turtle and Man*, 19.

2. A silver soup tureen in the shape of a green turtle, made by Paul de Lamerie in 1750 for an English plantation owner in Antigua, sold at Christie's London auction house in 1997 for £815,500 ($1,373,300). Sale 5825, lot 179, reported on Christie's website.

3. Ben Daley, Peter Griggs, and Helene Marsh, "Exploiting Marine Wildlife in Queensland: The Commercial Dugong and Marine Turtle Fisheries, 1847–1969," *Australian Economic History Review* 48, no. 3 (2008): 249.

4. Ibid., 246. The 1908 royal commission heard testimony suggesting that exploitation of the hawksbill turtle (*Eretmochelys imbricata*) for tortoise-shell exports was unsustainable, but recent price declines and reductions in registered tortoise-shell fishing vessels led the commission to focus on other fisheries. Ibid.

5. Ben Daley, "Changes in the Great Barrier Reef since European Settlement: Implications for Contemporary Management" (PhD diss., James Cook University, 2005), 389.

6. H. Robert Bustard, *Sea Turtles: Natural History and Conservation* (London: Collins, 1972), 154–55.

7. Daley, Griggs, and Marsh, "Exploiting Marine Wildlife in Queensland," 253.

8. Anthony Musgrave and Gilbert P. Whitley, "From Sea to Soup: An Account of the Turtles of North-West Islet," *Australian Museum Magazine* 2 (1926): 331–36.

9. Bustard, *Sea Turtles*, 156.

10. Musgrave and Whitley, "From Sea to Soup," 336.

11. Charles Barrett, "The Great Barrier Reef and Its Isles: The Wonder and Mystery of Australia's World-Famous Geographical Feature," *National Geographic Magazine* 58, no. 3 (1930): 375.

12. Celmara Pocock, "Turtle Riding on the Great Barrier Reef," *Society & Animals* 14, no. 2 (2006): 129–58.

13. F. W. Moorhouse, "Notes on the Green Turtle (*Chelonia mydas*)," in *Reports of the Great Barrier Reef Committee* 4, pt. 1, no.1 (1931): 19 (emphasis in original). Hereafter cited by page number in the text.

14. Bustard, *Sea Turtles*, 160–61.

15. Ibid.

16. Daley, Griggs, and Marsh, "Exploiting Marine Wildlife in Queensland," 255.

17. Ibid.

18. Bustard, *Sea Turtles*, 162–65.

19. Daley, *Changes in the Great Barrier Reef*, 258.

20. Bustard, *Sea Turtles*, 163; cf. Daley, *Changes in the Great Barrier Reef*, 408.

21. Daley, Griggs, and Marsh, "Exploiting Marine Wildlife in Queensland," 259.

22. Bustard, *Sea Turtles*, 165.

23. Ibid. The gazetted order of September 7, 1950, was repealed in 1958 and a new order enacted allowing commercial fishermen to capture green turtles in Queensland waters north of latitude 15 degrees south. Daley, Griggs, and Marsh, "Exploiting Marine Wildlife in Queensland," 259, citing Colin J. Limpus et al., "The green turtle, *Chelonia mydas*, Population of Raine Island and the Northern Great Barrier Reef, 1843–2001," *Memoirs of the Queensland Museum* 49, no. 1 (2003): 366–70.

24. Daley, *Changes in the Great Barrier Reef*, 406; Daley, Griggs, and Marsh, "Exploiting Marine Wildlife in Queensland," 258.

25. Daley, Griggs, and Marsh, "Exploiting Marine Wildlife in Queensland," 259.

26. Bustard, *Sea Turtles*, 166–67. Bustard suggests the 1968 legislation was based on his recommendation to the chief inspector of fisheries for Queensland

that all sea turtles be protected until such time as closed-cycle turtle farming was feasible and could be licensed by the government.

27. Daley, *Changes in the Great Barrier Reef*, 409.

28. Barrett, "Great Barrier Reef," 360, 375. The practice was started by naturalists who were visiting Masthead Island in the early 1900s, as illustrated by a photograph by Robert Etheridge, paleontologist and director of the Australian Museum. See also Daley, *Changes in the Great Barrier Reef*, 411, fig. 7.19.

29. Daley, *Changes in the Great Barrier Reef*, 410; Musgrave and Whitley, "From Sea to Soup," 336.

30. Barrett, "Great Barrier Reef," 360, 375.

31. Daley, *Changes in the Great Barrier Reef*, 415.

Chapter 3: The Turtle Islands of Sarawak

1. John R. Hendrickson, "The Green Sea Turtle, *Chelonia mydas* (Linn.), in Malaya and Sarawak," *Proceedings of the Zoological Society of London* 130, no. 4 (1958): 458 (quoting Keppel's report of the 1847 expedition of HMS *Dido*). See also Parsons, *Green Turtle and Man*, 61. The rajah Muda Hassim was the brother of the reigning sultan and heir presumptive to the throne of Borneo.

2. Hendrickson, "Green Sea Turtle in Malaya," 458–59.

3. Robert W. C. Shelford, *A Naturalist in Borneo* (London: T. Fisher Unwin, 1916), 301. Robert Shelford served as curator of the Sarawak Museum from 1897 to 1905 and experienced firsthand how seriously the turtle celebrations were taken. He brought a friend and fellow naturalist to Satang Island intending to make some observations and collect specimens but was told to keep out. Luckily, his companion knew a few words of Malay and was able to say "prenta," the Malay word for government. They were given permission to land on the far side of the island and collect invertebrates in the reef. Shelford described sea turtles in his 1916 book, but he did not make the detailed observations that later curators would make, especially Tom Harrisson.

4. Ibid., 110.

5. Tom Harrisson, "The Edible Turtle (*Chelonia mydas*) in Borneo, 11: West Borneo Numbers—The Downward Trend," *Sarawak Museum Journal*, n.s., 10 (1962): 614–23, and "Notes on Marine Turtles, 18: A Report on the Sarawak Turtle Industry (1966) with Recommendations for the Future," *Sarawak Museum Journal* 15, nos. 30–32 (1967): 424–36. Edward Banks, who became the museum's curator in 1925, said two million were usually taken in the 1930s. Banks, "The Breeding of the Edible Turtle (*Chelonia mydas*)," *Sarawak Museum Journal* 44, no. 4 (1937): 530.

6. Tom Harrisson, "The Edible Turtle (*Chelonia mydas*) in Borneo, 5: Tagging Turtles and Why," *Sarawak Museum Journal* 7, no. 8 (1956): 505.

7. Edward Banks, *A Naturalist in Sarawak* (Kuching, Sarawak: Kuching Press, 1949), 15. Banks noted that the fluctuations in number of nests laid indicated by the records suggested considerable variation in the number of eggs laid in a given year, citing his earlier paper, "Breeding of the Edible Turtle," 523–32.

8. Banks, *Naturalist in Sarawak*, 15.

9. Ibid. Banks also noted that the Malay chiefs were aware that only approximately 2–3 percent of the hatchlings survived once they reached the ocean and believed it would be wasteful to let the eggs hatch. Banks, "Breeding of the Edible Turtle," 532.

10. Hendrickson, "Green Turtle in Malaya," 459–60.

11. Lucas Chin, "Obituaries and Memorial Statements: Tom Harrisson and the Green Turtles of Sarawak," *Borneo Journal* 8, no. 2 (1976): 63–69.

12. Heimann, *Most Offending Soul Alive*, 22–24. Although not yet twenty-one, Harrisson had already published notes and articles in British journals, including *British Birds*, the *Journal of Ecology*, and *Nature*, Britain's most widely read scientific periodical. Ibid., 22.

13. For Harrisson's personal account of the South Pacific expedition, see his *Living among Cannibals* (London: Harrop, 1943).

14. Tom Harrisson, *Borneo Jungle: An Account of the Oxford Expedition to Sarawak* (London: L. Drummond, 1938).

15. Heimann, *Most Offending Soul Alive.*

16. Harrisson, "Notes on Marine Turtles, 18: Recommendations for the Future."

17. Harrisson, "Tagging Turtles (and Why)."

18. Heimann, *Most Offending Soul Alive*, 17. The British bird conservation movement in which the young Harrisson participated and the resulting British legislation inspired the formation of international conservation institutions, including the International Union for the Protection of Nature, which was formed in 1948. Mark V. Barrows, Jr., *Nature's Ghosts: Confronting Extinction from the Age of Jefferson to the Age of Ecology* (Chicago: University of Chicago Press, 2009), 310–12.

19. Harrisson, "Tagging Turtles (and Why)," 507.

20. Thomas P. Rebel, *Sea Turtles and the Turtle Industry of the West Indies, Florida, and the Gulf of Mexico* (Coral Gables: Univ. of Miami Press, 1974). The book's annotated bibliography includes twenty-four publications by P. E. P. Deraniyagala from 1930 to 1964, with nineteen published before 1949. Harrisson would have seen the book's first edition, prepared by Robert Ingle and F. G. Walton Smith, when Ingle was chief of Florida's Bureau of Marine Science and

Technology, and Smith was professor of oceanography at the University of Miami. Walton Smith's recommendations in 1950 for the rational exploitation of green turtles included discouraging export, a seventy-five-pound minimum size, and an experimental turtle farm and hatchery to study reproduction, among others.

21. Moorhouse wrote that "the necessity of forming large compounds in which to hold baby turtles until such time as they can fend for themselves is patent." "Notes on Green Turtles," 18.

22. Harrisson, "The Present and Future of the Green Turtle," *Oryx* 6, no. 5 (1962): 265–69, describing starting his hatchery "from scratch" in 1948. In a May 1964 letter to *The New Scientist* preserved in the Harrisson Papers, Harrisson pointed out that his turtle hatchery in Sarawak predated the leatherback hatchery Hendrickson started in 1961 in eastern (peninsular) Malaya.

23. Tom Harrisson, "The Sarawak Turtle Islands' 'Semah,'" *Journal of the Malayan Branch of the Royal Asiatic Society* 23, no. 3 (1950): 105–26.

24. Tom Harrisson, "The Edible Turtle (*Chelonia mydas*) in Borneo, 4: Growing Turtles and Growing Problems." *Sarawak Museum Journal* 7, no. 7 (1956): 235.

25. Ibid.

26. Harrisson, "Breeding of the Edible Turtle," *Nature* 169, no. 4292 (1952): 198.

27. Carla Kishinami, "John Roscoe Hendrickson Biography: A Daughter's Memories," Intercultural Center for the Study of Deserts and Oceans, www .cedointercultural.org/JRHtext.htm, accessed June 4, 2011; David W. Owens, "John Roscoe Hendrickson, 1921–2002," *Marine Turtle Newsletter* 99 (2003): 1–3.

28. Harrisson, "Tagging Turtles (and Why)," 509 (on the difficulties they had from 1950 to 1953 when the cow ear fold-over tag and spring pincers were devised); Hendrickson, "Green Sea Turtle in Malaya," 487–88; Heimann, *Most Offending Soul Alive*, 285–87; telephone interview with Judith Heimann, Nov. 18, 2009.

29. Harrisson, "Tagging Turtles (and Why)," 508–9.

30. Heimann, *Most Offending Soul Alive*, 285.

31. Telephone interview with Heimann, Nov. 18, 2009, in which she consulted the transcribed notes of her 1993 interview of John Hendrickson.

32. Hendrickson, "Green Sea Turtle in Malaya," 488; Harrisson, "Tagging Turtles (and Why)," 510.

33. Hendrickson, "Green Sea Turtle in Malaya," 488; Harrisson, "Tagging Turtles (and Why)," 510.

34. Telephone interview with Heimann, Nov. 18, 2009.

35. Hendrickson, "Green Sea Turtle in Malaya," 519.

36. Ibid., 489, 503, 497.

37. Ibid., 517.

38. Ibid., 485. In "Tagging Turtles (and Why)," 511, 509, Harrisson reported that he found the local atmosphere of goodwill absent, in part because the research had turned the island into a laboratory.

39. Harrisson, "Tagging Turtles (and Why)," 511.

40. The meeting in San Francisco was in July 1955. In their article "The Ecology and Migration of Sea Turtles, 2: Results of Field Work in Costa Rica, 1955," *American Museum Novitates*, no. 1835 (1957): 4, Archie Carr and Leonard Giovannoli refer to a paper by "J. R. Hendrickson" read at the meeting.

41. Carr and Caldwell, "The Ecology and Migrations of Sea Turtles, 1: Results of Field Work in Florida, 1955," 3.

42. Harrisson, "Tagging Turtles (and Why)," 509; Harrisson, "Tagging Green Turtles, 1951–56," *Nature* 178, no. 1475 (1956): 1479.

43. Harrisson, "Tagging Turtles (and Why)," 509. The emphasized phrase (which is in the original) presumably refers to all the help Hendrickson got from the Turtle Board staffers, who were Harrisson's employees.

44. Harrisson's conversation with Hendrickson about attribution probably occurred after October 1, 1956, the day Harrisson submitted the *Nature* letter, because after the confrontation, Harrisson would have acknowledged Hendrickson, albeit grudgingly. The dissension appears to have been over Harrison's second note on the "edible turtle." In this note, on copulation, Harrisson describes having observed, from a canoe, two turtles copulating. Hendrickson had actually made these observations, which were described in detail in the field notes Harrisson had asked him to leave with him. Tom Harrisson, "The Edible Turtle (*Chelonia mydas*) in Borneo, 2: Copulation," *Sarawak Museum Journal* 6, no. 4 (1954): 126–28.

45. Heimann, *Most Offending Soul Alive*, 286.

46. The *Sarawak Museum Journal* note has a postscript updating the return numbers as of October 1, 1956. Harrisson's letter to *Nature*, "Tagging Green Turtles, 1951–56," is also dated October 1.

47. Heimann, *Most Offending Soul Alive*, 287.

48. Ibid., 349.

49. Archie Carr to Tom Harrisson, March 7, 1956, box 19, Carr Papers, University of Florida Libraries, Gainesville.

50. Telephone interview with Heimann, Nov. 18, 2009.

51. Harrisson to Carr, June 5, 1956, box 19, Carr Papers.

52. Carr to Harrisson, Aug. 25, 1956, Harrisson Papers, University of Florida Libraries, Gainesville. See also Tom Harrisson, "Tagging Green Turtles, 1951–56";

Carr and Giovannoli, "Costa Rica, 1955," 4 (which describes applying the new tags in last four days of the field season).

53. Carr, *Handbook of Turtles*, 354–56.

Chapter 4: The Gifted Navigators

1. Carr, *Windward Road* (1956), 237.

2. Carr, "The Passing of the Fleet," *American Institute of Biological Sciences Bulletin* 4 (1954): 17.

3. Carr, *Handbook of Turtles*, 355.

4. Daley, *Changes in Great Barrier Reef*, 406–8.

5. During World War II, the *National Geographic* published a photographic essay on the fishery and its connection to Key West. See David D. Duncan, "Capturing Giant Turtles in the Caribbean," *National Geographic Magazine* 84, no. 2 (1943): 177–90. In his biography of Carr, Frederick Davis concludes that Carr saw this article while he was deciding to focus his research on sea turtles. Davis, *Man Who Saved Sea Turtles*, 59.

6. The Turtle Bogue (Tortuguero) fishery did not exist at the time of the Texas green turtle canneries and the decline of the green turtles in Texas waters. The Miskito Indians along the Caribbean coast subsistence fishery and the Cayman Island export fishery were probably the only drains on the population.

7. Carr, "Passing of the Fleet," 17.

8. Ibid.

9. Ibid., 19.

10. Carr included the 1954 AIBS address as the last chapter of *The Windward Road* (1956). According to his preface to the 1979 reissue, Carr published *The Windward Road*, a compilation of his natural history writings, to improve the prospects of the Caribbean green turtle by publicizing its plight. One chapter of *The Windward Road*, "The Black Beach," appeared in the July 1955 *Mademoiselle* and won the O. Henry Award for short stories even though it was nonfiction. *The Windward Road* won the John Burroughs Medal from the American Museum of Natural History for nature writing in 1957. The book led to the creation of the Brotherhood of the Green Turtle, forerunner to the Caribbean Conservation Corporation, as well as to ongoing support for Carr's research and conservation work. Carr's early mentor, Thomas Barbour, from Harvard's Museum of Comparative Zoology, had also written natural history essays for popular journals; and his evident enjoyment of this work and the additional income he derived from these articles may have inspired Carr to do the same. Davis, *Man Who Saved Sea Turtles*, 52. Carr dedicated *The Windward Road* to the memory of Thomas Barbour.

11. Carr, *Handbook of Turtles*, 345–57.

12. Ibid.

13. Ehrenfeld, *Biological Conservation*, 214.

14. Davis, *Man Who Saved Sea Turtles*, 123.

15. Carr and Giovannoli, "Costa Rica, 1955," 18 (referring readers to Carr's *Windward Road* for the story of how Carr learned the Caymanian turtles captains' "folk zoology").

16. Carr, *Windward Road* (1956), 11, 26.

17. Carr and Caldwell, "Florida, 1955," 3; Davis, *Man Who Saved Sea Turtles*, 120.

18. Carr, *Windward Road* (1956), 28–29; Archie Carr to Tom Harrisson, March 7, 1956, box 19, Carr Papers; Harrisson to Carr, June 5, 1956, box 19, Carr Papers; Harrisson, "Tagging Green Turtles, 1951–56."

19. Carr and Giovannoli, "Costa Rica, 1955," 11.

20. Ibid. Carr was a research associate of the American Museum of Natural History.

21. Davis, *Man Who Saved Sea Turtles*, 142. Jim Oliver was one of the first members of the board of the Caribbean Conservation Corporation, the nonprofit organization that was formed to help Carr's research.

22. Carr and Giovannoli, "Costa Rica, 1955," 22, table 6. Ebanks also caught a turtle bearing only tag holes at the Miskito cays. Giovannoli examined this turtle at Key West and verified that it had been one of the last to be tagged with the ineffective shell tag during the 1955 field season.

23. Carr, *Windward Road* (1979), 214.

24. Ibid., 232–35.

25. John R. Hendrickson to Archie Carr, Jan. 14, 1957, Carr Papers. Hendrickson's letter arrived while Carr was teaching at the University of Costa Rica. Davis, *Man Who Saved Sea Turtles*, 128. Harrisson's letter in *Nature*, "Tagging Green Turtles, 1951–56," included a reference to a letter from Archie Carr telling him that the flipper tag had been adopted in Tortuguero.

26. Archie Carr, "The Zoogeography and Migrations of Sea Turtles," *Yearbook of the American Philosophical Society*, 1954, 140.

27. Carr and Caldwell, "The Ecology and Migrations of Sea Turtles, 1: Results of Field Work in Florida, 1955." Caldwell may have told Carr that Harrisson had given the paper or assumed it was Harrisson because of the latter's recent letter to *Nature* (Harrisson, "Tagging Green Turtles, 1951–56").

28. Hendrickson to Carr, Jan. 14, 1957, Carr Papers.

29. Carr to Hendrickson, March 28, 1957, Carr Papers.

Chapter 5: The Geography of Turtle Soup

1. James J. Parsons, "The Miskito Pine Savanna of Nicaragua and Honduras," *Annals of the Association of American Geographers* 45, no. 1 (1955): 46, 51. In *High Jungles and Low*, Carr speculated that hurricane blow-downs had led to single-species patches of vegetation behind Pearl Lagoon, in eastern Nicaragua.

2. Davis, *Man Who Saved Sea Turtles*, 83.

3. James J. Parsons, "English Speaking Settlement of the Western Caribbean," *Yearbook of the Association of Pacific Coast Geographers* 16 (1954): 3–16; Parsons, "Miskito Pine Savanna."

4. James J. Parsons to Archie Carr, Feb. 11, 1959, and March 5, 1959, Carr Papers, Series 2. In his March 5 letter, Parsons told Carr that Hendrickson had promised to send him two forthcoming papers on Malaya turtles. The two Hendrickson papers appeared in vol. 26 of *Bulletin of Raffles Museum* (1961) but are not cited in Parsons' bibliography, suggesting that they were not published until after Parsons' book, *The Green Turtle and Man*, went to press; their omission may have been one issue nettling Hendrickson when he wrote his review of the book.

Parsons also told Carr that he was still looking for a paper on Thailand from the Pacific Science Congress held in Bangkok in November 1957: "I am not satisfied with the evidence on Buddhist prejudices and am following up a few leads." He needed information about the coast of Siam, where he was fairly sure eggs were a big business but, as in Malaya, the meat did not appear to be eaten. Parsons to Carr, Feb. 11, 1959. The Thailand paper Parsons was looking for was one presented by Commander A. Penyapol, "A Preliminary Study of the Sea Turtles in the Gulf of Thailand" (published separately by the Hydrographic Department, Royal Thai Navy). Parsons used it in his account of green turtle use in Siam. Parsons, *Green Turtle and Man*, 56–57.

5. Parsons, *Green Turtle and Man*, 59.

6. Ibid., 18. See also Roger Huxley, "Historical Overview of Marine Turtle Exploitation, Ascension Island, South Atlantic," *Marine Turtle Newsletter* 84 (1999): 7–9.

7. Parsons, *Green Turtle and Man*, 18.

8. Ibid., 17–18.

9. Parsons to Carr, March 25, 1959, Carr Papers.

10. Tom Harrisson, "The Edible Turtle (Chelonia mydas) in Borneo, 6: Semah Ceremonies, 1949–1958," *Sarawak Museum Journal* 8, no. 11 (1958): 482–86.

11. Parsons, *Green Turtle and Man*, 9, citing Hendrickson, "Green Sea Turtle in Malaya," 457–58.

12. Archie Carr to Lewis Haines, April 8, 1959; Carr to M. Graham Nutting, May 21, 1959; and Nutting to Carr, May 26, 1959, all in the Carr Papers.

Parsons' book combined the historical focus of his mentor, Carl Sauer, with his own growing concern over contemporary environmental problems. Among the many favorable reviews of the book as a work of cultural biogeography, there was never a review in the premier journal of the discipline, *Annals of the Association of American Geographers*. Bret Wallach, "In Memoriam: James J. Parsons, 1915–1997," *Annals of the Association of American Geographers* 88, no. 2 (1998): 320.

13. Archie Carr, preface to the 1979 edition of *The Windward Road*, xv.

14. This account of the formation of the Brotherhood of the Green Turtle is from the foreword Joshua Powers wrote for the 1979 edition of *The Windward Road*. Carr also gives an account in his preface. Carr, *Windward Road* (1979 ed.), xiv–xv.

15. Davis, *Man Who Saved Sea Turtles*, 164–66 (citing Carr's reports as Caribbean Conservation Corporation technical director). Carr gives more information about Operation Green Turtle in his book, *So Excellent a Fishe: A Natural History of Sea Turtles* (Garden City, NY: Natural History Press, 1967), and in the 1979 edition of *The Windward Road*. In the latter, he acknowledges the mistakes the participants made because they believed that the turtles reached sexual maturity at five or six years, and that the hatchlings might imprint somehow on their release sites rather than return to Tortuguero to breed. By the sixth year (about 1966), Operation Green Turtle was translocating eggs instead of hatchlings, based on the belief that imprinting occurred during embryonic development. Carr, *Windward Road* (1979 ed.), xvi–xvii.

16. Carr, foreword to Parsons, *Green Turtle and Man*, v.

17. Archie Carr and Larry Ogren, "The Ecology and Migration of Sea Turtles, 4: The Green Turtle in the Caribbean Sea," *Bulletin of the American Museum of Natural History* 121, no. 1 (1960): plate 1. Parsons had asked Carr for photos for the book; because his research was a "library job," he had none of his own. Carr promised pictures several times, including some of the Cayman turtle schooners, but ended up giving Parsons several from the research at Tortuguero. Carr's 1957 *American Museum Novitates* article with Giovannoli, "Results of Field Work in Costa Rica, 1955," had photos Giovannoli had taken of turtles in a schooner's hold and in kraals, but Carr did not give these to Parsons in 1959.

18. John R. Hendrickson, Review of *The Green Turtle and Man*, by James J. Parsons, *Science* 140 (May 24, 1963): 885.

19. Parsons was president of the Association of Pacific Coast Geographers and later of the American Association of Geographers, serving twice as chair of the geography department at the University of California–Berkeley. Wallach, "In Memoriam."

20. As we have seen, Parsons used the map in one endpaper; the other endpaper was his own map of the nesting and feeding grounds around the world.

21. Carr and Ogren, "Green Turtle in the Caribbean Sea," 28. It is not clear from the letters whether Carr saw the "slip of some sort" when he reviewed Hendrickson's manuscript in March 1957, but as noted in chapter 4, Carr had told Hendrickson he was disappointed that Hendrickson's research had shed no light on the issue of long-range migration. When the Carr and Ogren paper came out, Hendrickson took mild exception to Carr's characterization of his conclusion on site fidelity but told Carr these were only "personal comments for your own enlightenment." He presumed Carr could have judged from the sentence following the one that he quoted that Hendrickson was referring to consistency over a lifetime because the evidence Hendrickson had found was strong only for consistency within one breeding season. "I am unable, by the way, to agree with you that my data show by inference that the turtles return to nest on the island where they themselves had hatched. I cannot eliminate the possibility of a general "pool" of turtles in the South China Sea, with each annual increment of newly-matured individuals following a drift or migration of older, experienced adults to breeding beaches, then fixing their location sense and its attendant "homing" instinct on a substrate of individual adult experience rather than racial memory or memory of infant experience. It would indeed be a pleasure to have an opportunity to discuss some of these questions with you in person some time." In a postscript connected by a penciled-in arrow to the above paragraph, Hendrickson added, "I am still stumped by the problem of marking hatchlings so they will be recognizable when adults—have you any ideas? JRH." Hendrickson to Carr, April 12, 1961, Carr Papers, Series 2. Hendrickson made no comment on the plastic disks applied to 20 percent of the hatchlings delivered to distant sites under Operation Green Turtle, as described in Parsons' photo essay in the concluding chapter. Hendrickson wrote again two weeks later and softened his tone, saying, "We obviously are not going to disagree on any of the substantiated aspects of the Sea Turtle problem. My 'pool' of South China Turtles is actually not a real belief on my part either. It is only a concept which I cannot as yet exclude from my thinking." Hendrickson to Carr, May 2, 1961, Carr Papers.

22. Carr and Ogren, "Green Turtle in the Caribbean Sea," 29: "In spite of . . . the lack of statistical support for the belief, we are convinced, as we have said, that the renesting returns alone indicate a marked homing urge and ability in the female Caribbean green turtle."

23. Parsons to Carr, April 12, 1961, Carr Papers.

24. Carr, *The Reptiles*.

Chapter 6: A Turtle Flap in London

1. Tom Harrisson, "The Edible Turtle (*Chelonia mydas*) in Borneo, 9: Some New Hatching Observations," *Sarawak Museum Journal* 10, nos. 17–18 (1961): 299. See also Harrisson, "West Borneo Numbers," 614–23.

2. Tom Harrisson, "Present and Future of the Green Turtle," 265–69.

3. Ibid. Harrisson's remark about the name *Mydas* might have been a reference to the legend in Greek mythology of King Midas, who turned everything he touched into gold.

4. Heimann, *Most Offending Soul Alive*, 337. See Barbara Harrisson's book *Orang-utan*.

5. Heimann, *Most Offending Soul Alive*, 129, 162, 391–92.

6. Mark V. Barrows, Jr., *Nature's Ghosts*, 315–16.

7. Ibid., 311. At the time of its founding, the IUCN was called the International Union for Nature Preservation.

8. Ibid., 315–17.

9. Davis, *Man Who Saved Sea Turtles*, 181–83. Carr told IUCN officials that he would rather see Costa Rica adopt the park at Tortuguero as an act of national sovereignty over its natural resources than at the behest of an international campaign.

10. *Survival Species Memorandum, No. 14* (London: Fauna Preservation Society, Zoological Society of London, Sept. 24, 1963), in Harrisson Papers, IUCN Series, box 7. See also Barrows, *Nature's Ghosts*, 315–17. In 1964, biologists at the US Bureau of Sport Fisheries and Wildlife created a committee on wildlife which published in August 1964 the first Red Book of threatened species, following the model of IUCN's Survival Service Committee. Yaffee, *Prohibitive Policy*, 34–35.

11. Davis, *Man Who Saved Sea Turtles*, 169–70.

12. Barbara Harrisson, "International Proposals to Regulate Trade in Non-human Primates," *Primates* 13, no. 1 (March 1972): 111–14.

13. Ehrenfeld, *Biological Conservation*, 109.

14. Tom Harrisson, "Present and Future of the Green Turtle," 265–69.

15. Tom Harrisson to David Astor, 1963, Harrisson Papers, Correspondence Series.

16. Tom Harrisson, "Must the Turtle Die?" *Sunday Times* (London) *Weekly Review*, June 14, 1964, 29.

17. LaCroix to IUCN, June 1964, Harrisson Papers, IUCN Series. Harrisson somewhat retracted his insinuations, however, when he explained his meaning to David Astor. Harrisson to Astor, July 1964, Harrisson Papers, Correspondence Series.

18. Davidson, *Fire in the Turtle House*, 33.

19. John R. Hendrickson and E. Balasingham, "Nesting Beach Preferences of Malayan Sea Turtles," *Bulletin of the National Museum of Singapore* 33, no. 10 (1966): 69–76.

20. Kishinami, "John Roscoe Hendrickson Biography."

21. John L. Culliney, *Islands in a Far Sea* (Honolulu: University of Hawaii Press, 2006), 118–19.

22. Daniel Kaha ʻulelio, *Ka ʻOihana Lawaiʻa: Hawaiian Fishing Traditions* (Honolulu: Bishop Museum Press, 2006), 239. There is disagreement among historians on whether turtles were eaten by the common people or reserved for the nobility.

23. Hendrickson, "IUCN Report on Hawaiian Marine Turtles," Harrisson Papers, IUCN Series.

24. Ernest S. Reece to Tom Harrisson, Aug., 1965, Harrisson Papers, Correspondence Series.

Chapter 7: The Buffalo of the Sea

1. Carr, *The Reptiles*, 175.

2. Carr, *Handbook of Turtles*, 357.

3. Carr, *Windward Road* (1979 ed.). In this edition (p. 252), Carr included a photograph of young farm-reared green turtles at Torres Strait, Australia. Carr took this photograph in October 1973 when he visited the farms on behalf of the Australian government's inquiry into the turtle farms.

4. David K. Caldwell and Archie Carr, "Status of the Sea Turtle Fishery in Florida," in *Transactions of the Twenty-Second North American Wildlife Conference (March 4–6, 1957)*, (Washington, DC: Wildlife Management Institute, 1957), 460. The first North American Wildlife Conference had been called by Franklin D. Roosevelt in February 1936 and included a session entitled "The Problem of Vanishing Species." The proceedings were published as a US congressional committee print, for the 74th Congress, 2nd session.

5. Caldwell and Carr, "Sea Turtle Fishery in Florida," 461–62.

6. Ibid., 462–63.

7. Harold Coolidge was the vice president of IUCN from 1948 to 1954, then its president, and was the founding chair of IUCN's Survival Service Commission. See James L. Aldrich and Anne M. Blackburn, "A Tribute to Harold J. Coolidge," *Environmentalist* 5, no. 2 (1985): 83–84; and Lee M. Talbot, "Dedication to Dr. Harold J. Coolidge," *Environmentalist* 2, no. 4 (1982): 281–82. Coolidge co-organized the Pacific Science Congress every four years. He may have met John Hendrickson at the 1953 Congress in Manila. At the 1957 congress in

Bangkok, Coolidge recommended that countries of Southeast Asia become involved in IUCN so that they could coordinate their conservation efforts. At that session, Tom Harrisson described the status of conservation in Borneo, noting that uncontrolled trade for zoos and pets was threatening many native species. The one exception was the green turtle because he controlled the level of egg collecting. Harrisson, "Conservation in the Island of Borneo," 12.

8. Archie Carr, "The Navigation of the Green Turtle," *Scientific American* 212, no. 5 (1965): 84.

9. Carr and Ogren, "Ecology and Migration of Sea Turtles, 4: The Green Turtle in the Caribbean Sea," 23–24. Carr and Ogren described informal tests of the sea-finding ability of hatchlings and female nesters under different conditions of beach gradient, distance from water, light intensity, and so on. Ibid., 29–46.

10. Carr, *So Excellent a Fishe*, 87–88. Carr noted that they needed "more searching tests" to determine what turtles see and how they react to light in finding the sea.

11. David Ehrenfeld was a PhD student of Carr's in the mid-1960s working on the behavioral ecology of green turtles and a National Institutes of Health fellow during the 1966 field work. He published the results with Arthur L. Koch. See David W. Ehrenfeld and A. L. Koch. "Visual Accommodation in the Green Turtle," *Science* 155 (Feb. 17, 1967): 827–28. Ehrenfeld went on to write extensively on the scientific and philosophical aspects of biological conservation, including the first textbook on the subject, drawn from his courses at Barnard College. He cautioned against scientism, saying that "science will need careful guidance and supervision from other disciplines" if it is to play a positive role in the future of biological conservation.

12. Carr described this work in *So Excellent a Fishe,* 89–90.

13. Several of these students went on to careers that allowed them to continue to contribute to Carr's work in sea turtle conservation. Larry Ogren went to work for the National Marine Fisheries Service; Harold Hirth, to the University of Utah, with stints at the UN's Food and Agriculture Organization and Fulbright fellowships to Qatar. Howard Campbell went to work at the Department of Interior's Office of Endangered Species. David Ehrenfeld taught at Barnard College and then Rutgers University, where he authored the first textbook in biological conservation and founded the first scientific journal in that field, *Conservation Biology*. F. Wayne King was conservation director for the New York Zoological Society, where he actively campaigned to list Florida's endangered reptiles, including the American alligator and the green turtle, and held leadership positions in IUCN's Species Survival Commission.

14. Carr had regular columns in *Animal Kingdom* and other magazines. He told the story of his research on migration in Carr, "Navigation of the Green Turtle."

15. Archie Carr, "Alligators: Dragons in Distress," *National Geographic* 131, no. 1 (1967): 133.

16. Robert E. Schroeder and William A. Starck II, "Diving at Night on a Coral Reef," *National Geographic* 125, no. 1 (1964): 128–54; Robert E. Schroeder, *Something Rich and Strange* (New York: Harper & Row, 1965). Schroeder and Starck invented the bang stick to deter shark attacks.

17. This account is based on three sources: Fosdick and Fosdick, *Last Chance Lost?* 8; Bustard, *Sea Turtles*, 183; and Barbara C. Daley, "Why We Still Enjoy a Plate of Turtle in Cayman Today," *Cayman Net News*, Aug. 17, 2007. Publications from Schroeder's doctoral research include Robert E. Schroeder and W. H. Leigh, "The Life History of *Ascocotyle pachycystis* sp.n., a Trematode (Digenea: Heterophyidae) from the Raccoon in South Florida," *Journal of Parasitology* 51, no. 4 (1965): 594–99. Both Fosdick and Fosdick and Bustard quote Schroeder as saying that Wayne King was the graduate student at Miami who gave Schroeder his first batch of green turtles. King, however, did not recall having done so when interviewed by the author in Gainesville, Florida, on October 14, 2009.

18. Fosdick and Fosdick, *Last Chance Lost?* 9.

19. Ibid., 10–11. Jean Schroeder's photograph at Cape Sable was included in Schroeder, "Buffalo of the Sea," 178. A reporter's photo of the release appeared in Archie Carr, "Caribbean Green Turtle: Imperiled Gift of the Sea," *National Geographic* 131, no. 6 (1967): 882.

20. Schroeder, "Buffalo of the Sea," 176–83. His idea was that a rancher would raise hatchlings to a less vulnerable size, release them to fatten on undersea ranges of seagrass that were either enclosed or open, and then harvest a percentage of those when they returned to the nesting beach, selectively breeding turtles for rapid growth, large size, and good flavor.

21. Carr, "Caribbean Green Turtle," 886.

22. Ibid., 882. Nevertheless, Costa Rica's decrees in 1963 prompted Carr to go to the general assembly of IUCN in Nairobi, Kenya, to seek a resolution commending Costa Rica's action and to offer it as a model for other nations.

23. Archie Carr, "Great Reptiles, Great Enigmas," *Audubon*, March 1972, 25. This irritation is evident in Carr's letter to Polly Bergen, July 1972, Carr Papers, Series 2; and in his 1967 draft Red Data Book page for *Chelonia mydas*, Carr Papers.

24. This account of the discussions between Carr, Ehrenfeld, and the Schroeders is inferred from what each of them wrote almost contemporaneously. In

the book he was just finishing writing (*So Excellent a Fishe*, 229), Carr described the challenges any turtle farm would face and how the Schroeders had managed to avoid several of them. He had finished the final chapter, "Sea Turtles and the Future," in March 1966, according to Davis (*Man Who Saved Sea Turtles*, 171) and discussed turtle farming in the last pages of that chapter. Schroeder sketched out the idea of turtle ranching in "Buffalo of the Sea," 176–83. Ehrenfeld discussed green turtle farming in a textbook he developed from courses he began teaching after finishing his doctorate in zoology. Ehrenfeld, *Biological Conservation*, 196–199. Carr clarified his views on farming in an epilogue to the 1984 reissue of *So Excellent a Fishe*.

25. The Caribbean Conservation Commission built the Union Creek Experimental Green Turtle Culture Project with the help of the Bahamas National Trust, the Lerner Marine Laboratory of Bimini, the National Audubon Society, and the Morton Salt Company. Carr, *So Excellent a Fishe*, 234, 237.

26. Davis, *Man Who Saved Sea Turtles*, 167 (citing Carr, "Technical Director's Report to CCC, 1962–63"). Hurricane Flora hit the Bahamas in early October 1963 after lingering over Tobago and Cuba. It was the deadliest hurricane to hit the Caribbean in the twentieth century, with an estimated 8,000 lives lost.

27. Ibid., 167; Ehrenfeld, *Biological Conservation*, 197–99. A photo in Ehrenfeld's book (p. 197) shows Samuel Nixon holding a three-year old green turtle weighing sixty pounds at the Union Creek project.

28. Ehrenfeld, *Biological Conservation*, 198.

29. Robert E. Schroeder to H. Robert Bustard, May, 1969, quoted at length in Bustard, *Sea Turtles*, 183–86.

30. Carr, "Caribbean Green Turtle: Imperiled Gifts of the Sea." The original title had been "Gift from Turtle Bogue," and the final version when purchased by *National Geographic* was "Land of Turtle Mountain." Carr Papers, 1999 Addendum, box 5.

31. Ibid., 879.

32. Fosdick and Fosdick, *Last Chance Lost?* 19, 47; B. L. Lipp and R. L. Lipp to Robert E. Schroeder with copy to Archie Carr, June 25, 1966, Carr Papers, Series 2.

33. Fisher apparently learned the method from a friend who had visited Cornell University's agriculture research station. Fosdick and Fosdick, *Last Chance Lost?* 45–46.

34. Carr, "Caribbean Green Turtle," 887.

35. Fosdick and Fosdick, *Last Chance Lost?* 15, 19.

36. Ibid., 30.

37. Schroeder believed green turtles reached breeding age at six years and at the size of 200 pounds. R. E. Schroeder, "The Green Turtle: New Domestic Animal," undated report on Mariculture, Ltd., accompanying Schroeder to Carr, March 17, 1969, Carr Papers.

38. Bernard Nietschmann, *The Caribbean Edge: The Coming of Modern Times to Isolated People and Wildlife* (New York: Bobbs-Merrill, 1979), 207.

39. Schroeder to Carr, March 1969, Carr Papers, Series 2.

40. Schroeder to Carr, March 17, 1969, and Schroeder, "The Green Turtle: New Domestic Animal." Carr had sent a wire to the Costa Rican government and a copy to Nicaragua's institute of national development, urging them not to build the factories. With Carr's encouragement, the Caribbean Conservation Corporation's Billy Cruz went to San José shortly afterwards to urge the three countries to adopt a three-year moratorium on turtle hunting.

41. Schroeder to Carr, March 17, 1969, Carr Papers.

42. Nietschmann, *Caribbean Edge*, 210.

Chapter 8: Who Will Kill the Last Turtle?

1. James J. Parsons and Robert C. West, "The Topia Road: A Trans-Sierran Trail of Colonial Mexico," *Geographical Review* 31, no. 3 (1941): 406–13.

2. William M. Denevan, "Bernard Q. Nietschmann, 1941–2000: Mr. Barney, Geographer and Humanist," *Geographical Review* 92, no. 1 (2002): 104–9.

3. Parsons, *Green Turtle and Man*, 27.

4. Ibid., endleaf map. See also Archie Carr and Harold Hirth, "The Ecology and Migrations of Sea Turtles, 5: Comparative Features of Isolated Colonies," *American Museum Novitates*, no. 2091 (1962): 1–42.

5. Nietschmann, *Caribbean Edge*, 9.

6. This account of Nietschmann's visit to Tortuguero and conversation with Archie Carr is a reconstruction based on Nietschmann, *Caribbean Edge*, 8–9, 28, 61, 189, 197–201, 213–14; Carr, *So Excellent a Fishe* (1967), 41–72 (which Carr had just finished writing at the time of Nietschmann's visit); telephone interview with Judith (Nietschmann) Fitzpatrick (who traveled with Nietschmann to Costa Rica in 1967), Feb. 9, 2010; and email from William M. Denevan, Oct. 16, 2011 (Denevan was Nietschmann's advisor at the University of Wisconsin–Madison, and his own dissertation advisor at UC–Berkeley had been James J. Parsons). The occasion for Nietschmann's visit was a field trip with a summer ecology course at the Organization for Tropical Studies. Carr was very likely to have discussed green turtle ecology and his views on the farming issue with the visiting students. Nietschmann met several turtle fishermen from Nicaragua while at Tortuguero.

7. Archie Carr to Peter Scott, March, 1968, Carr Papers, Series 2.

8. Parsons, *Green Turtle and Man*, 23.

9. Ibid., 24.

10. Duncan, "Capturing Giant Turtles."

11. Nietschmann, *Caribbean Edge*, 210.

12. Bernard Nietschmann, *Between Land and Water: The Subsistence Ecology of the Miskito Indians, Eastern Nicaragua* (New York: Seminar Press, 1973), 38.

13. Carr, *So Excellent a Fishe* (1984 ed.), 58–59.

14. Ibid., 59, 60.

15. Davis, *Man Who Saved Sea Turtles*, 174.

16. See Matthiessen, "To the Miskito Bank"; Peter Matthiessen to Archie Carr, Feb. 28, 1964, Carr Papers, Series 2. Matthiessen went on to write *Far Tortuga* (1975), a fictional account of the last Cayman turtle schooner to convert to diesel and of the impact of economic and ecological changes upon the few remaining Cayman turtle fishermen. See John R. Cooley, "Waves of Change: Peter Matthiessen's Caribbean," *Environmental Review* 11, no. 3 (1987): 223–30.

17. Allie Ebanks to Archie Carr, June 28, 1959, reprinted in Carr, *So Excellent a Fishe* (1984 ed.), 62.

18. Ibid. Ebanks also wrote: "I saw an article while I was in Managua where that government and some people in Costa Rica along with your help are asking more consideration regarding the interference with laying turtles on [Tortuguero] beach."

19. David W. Ehrenfeld, telephone interview with author, Oct. 10, 2010.

20. Carr, *So Excellent a Fishe*, 234–38.

21. Parsons, *Green Turtle and Man*, 31; Cooley, "Waves of Change."

22. Nietschmann, "When the Turtle Collapses," 208.

23. Nietschmann, *Caribbean Edge*, 201–2.

24. Ibid.

25. Matthiessen, "To the Miskito Bank." See also Nietschmann, *Caribbean Edge*, 207, and Parsons, *Green Turtle and Man*, 31. In April 1965, Nicaragua declared a 200-nautical-mile "exclusive fishing zone" coterminous with an extended territorial sea. US Navy Judge Advocate General Corps, *Maritime Claims Reference Manual*. The decree was Executive Decree no. 1-L. See also the note on *Nicaragua v. Honduras* (ICJ 2007), *American Journal of International Law* 102, no. 1 (2008): 113–19 (evidence of Nicaragua's claim over marine resources).

26. Nietschmann, "When the Turtle Collapses," 34.

27. Ibid.

28. Ibid., 41, 42.

Chapter 9: Red Data for the Green Turtle

1. Archie Carr to Peter Scott, Feb. 16, 1966, Carr Papers, Series 2, box 22.
2. Davis, *Man Who Saved Sea Turtles*, 174.
3. Carr, *So Excellent a Fishe*, 226, 228–29.
4. Carr to Scott, Feb. 16, 1966.
5. Draft Red Data Book page for *Chelonia mydas*, May 1968, Harrisson Papers, IUCN Series, box 7.
6. When the Survival Service Commission reviewed the draft Red Data Book pages compiled on reptiles at its meeting in May 1968, Harrisson made sure to add, "By 1966, hatch rate was up to 80% and over," citing "Harrisson in litt. May 17, 1968." R. E. Honegger, "Rare and Endangered World Amphibians and Reptiles, July 1968," in Harrisson Papers, IUCN Series, box 7.
7. Harrisson made sure the pages for the green turtle, which were based on Carr's submission, included this language under "protective measures proposed." Harrisson Papers, IUCN Series, box 7. René E. Honegger, curator of reptiles at the Zurich Zoo, prepared the first Red Data Book pages for reptiles and amphibians in 1967 based on Carr's submission (and Harrisson's insertions). Davis, *Man Who Saved Sea Turtles*, 175.
8. Carr, "The Black Beach." Mrs. Ybarra was actually Juana Lopez, a resident of the village of Tortuguero, who helped Carr learn to locate nests of leatherback turtles and tried to sell him skins of an ocelot and a jaguar she had killed in her henhouse. Carr, *Windward Road* (1979 rev. ed.), photo opposite p. 69.
9. Hendrickson, "Green Sea Turtle in Malaya," 525–29; Carr, *So Excellent a Fishe* (1967), 229.
10. That Carr may have been trying to appeal to the FAO in recommending pilot turtle culture projects is evident in correspondence with Colin Holloway about finding funding agencies and tailoring the conservation program to meet the agencies' agendas. Carr to Holloway, May 3, 1968, Carr Papers, Series 3, box 28, folder 2.
11. Carr, *So Excellent a Fishe*, 229–32.
12. Davis, *Man Who Saved Sea Turtles*, 175.
13. The new draft also included a recommendation borrowed from the loggerhead page, that roadways and tourist hotels be located away from nesting beaches so that hatchlings are not disoriented by the lighting and fail to reach the sea. Honegger, final Red Data Book, volume on amphibians and reptiles (July 1968), Harrisson Papers, IUCN Series, box 7.
14. Correspondence between Holloway and Harrisson, 1968, Harrisson Papers, IUCN Series, box 7.

15. Harrisson to Holloway, May 14, 1968, Harrisson Papers, IUCN Series, box 7; Peter Charles Harold Pritchard, *Tales from the Thébaïde: Reflections of a Turtleman* (Malabar, FL: Krieger Publishing, 2007), 247; Carr to Holloway, May 3, 1968, Carr Papers, Series 3, box 28, folder 2. Twelve turtle specialists representing different regions attended: A. Carr (Caribbean and Ascension Island), P. C. H. Pritchard (Guyana and French Guiana), J. P. Schulz (Surinam), A. E. Montaya (Mexico), H. Hirth (Seychelles and Aldabra), George Hughes (South Africa), H. R. Bustard (Australia), J. R. Hendrickson (Hawaii), E. Balasingam (Malaysia), G. S. de Silva (Ceylon and Sabah), T. Harrisson (Sarawak and Philippines), and L. Brongersma (observer).

16. Carr to Harrisson, Dec. 18, 1968, Harrisson Papers, IUCN Series, box 7. In this letter, Carr wrote, "The guy seems constantly determined to forget the past and take a fresh start in every venture he makes into natural history."

17. Ibid.

18. H. R. Bustard to Harrisson, Dec. 4, 1968, Harrisson Papers, IUCN Series, box 7.

19. Davis, *Man Who Saved Sea Turtles*, 177; Carr to Holloway, April 15, 1969, Carr Papers, Series 3, box 28, folder 3.

20. Draft Notes on Working Meeting of Marine Turtle Specialists, March 10–13, 1969, Harrisson Papers, IUCN Series, box 7, 5.

21. Draft Red Data Book page for *Chelonia mydas*, May 1968, Harrisson Papers, IUCN Series, box 7.

22. Draft Notes on Working Meeting, Harrisson Papers, IUCN Series, box 7, 4.

23. Ibid.

24. For the seven-point program, see Tom Harrisson, "The Turtle Tragedy," *Oryx* 10, no. 2 (Sept. 1969): 112–15. In the opening paragraphs of this article, Harrisson recorded the statement adopted by the turtle group.

25. Draft Notes on Working Meeting, Harrisson Papers, IUCN Series, box 7, 4–5.

26. Ibid., 11.

27. H. Robert Bustard and K. P. Tognetti, "Green Sea Turtles: A Discrete Simulation of Density-Dependent Population Regulation," *Science* 163 (1969): 939–41.

28. In Harrisson Papers, IUCN Series, box 7: Holloway to Harrisson, March 19, 1969; Holloway to Leo Brongersma, April 9, 1969; Harrisson to Peter Pritchard, March 10, 1970. In Carr Papers, Series 3, box 28, folder 3: Carr to Holloway, April 15, 1969.

29. Carr to Holloway, May 3, 1968, Carr Papers, Series 3, box 28, folder 2.

30. George Hughes, telephone interview by the author, May 21, 2011.

31. Carr to Holloway, April 15, 1968, Carr Papers, Series 3, box 28, folder 3.

32. Bustard, *Sea Turtles.*

33. Holloway to Wolfgang Burhenne (IUCN, Commission on Legislation), April 2, 1969, Harrisson Papers, IUCN Series, box 7. Stanley de Silva was one of the least reluctant members of the specialists' group to urge legislative protection and is considered an unsung hero of the Indo-Pacific sea turtles. Sally Hopkins-Murphy, personal communication.

34. Holloway to Leo Brongersma, April 9, 1969, Carr Papers, Series 3, box 28, folder 3.

35. Carr to G. Yglesias, April 28, 1969, Carr Papers, Series 2, P. Scott folder (attached to letter to Peter Scott, April 28, 1969).

36. Tripartite Meeting about the Green Turtle, San José, Costa Rica, Sept. 30–31 and Oct. 1, 1969 (translated by Laura Chen Allen, March 2010), in author's possession.

37. Carr, "Great Reptiles, Great Enigmas," 29–30.

38. Ibid., 29–30. Carr cites Nietschmann's forthcoming article in *Human Ecology* ("Hunting and Fishing Focus among the Miskito Indians, Eastern Nicaragua," *Human Ecology* 1, no. 1 [1972]: 41–67), describing the impact of the factories on the Miskito's diet, when they sold the turtles to the factories instead of providing it for village subsistence.

39. Carr, "Great Reptiles, Great Enigmas," 30.

Chapter 10: Reptiles on the Red List

1. Archie Carr to Colin Holloway, April 15, 1969, Carr Papers, Series 3, box 28, folder 3.

2. Harrisson, "Turtle Tragedy," 112–15.

3. Heimann, *Most Offending Soul Alive,* 361.

4. Ibid., 365.

5. Harold F. Hirth, "Synopsis of the Biological Data on the Green Turtle, *Chelonia mydas* Linnaeus 1758," FAO Fisheries Synopsis No. 58 (Rome: UN Food and Agriculture Organization, 1971).

6. H. Robert Bustard, "Turtles and an Iguana in Fiji," *Oryx* 10 (1970): 317–22.

7. Holloway to Carr, March 19, 1969, Carr Papers, Series 3, box 28, folder 3.

8. "Revised Classification of Rare and Endangered Forms," Jan. 1969, enclosure in René Honneger to Marine Turtle Group, n.d., Carr Papers, Series 3, box 28, folder 4.

9. Honneger to Marine Turtle Group. Bustard also recommended changing the status of the loggerhead turtle from rare to depleted and the leatherback turtle from endangered to rare.

10. "Data Sheet for Green Turtle," enclosure in Honneger to Marine Turtle Group.

11. Udall's views on conservation at the time are presented in his book *The Quiet Crisis* (New York: Holt, Rinehart and Winston, 1963).

12. Barrows, *Nature's Ghosts*, 321.

13. Barrows, "Dragons in Distress," 279; Barrows, *Nature's Ghosts*, 322. The captive breeding facility received $350,000 in funds with the assistance of the ranking minority member of the Senate Appropriations Committee from South Dakota, who, after a staff member had been impressed by the sight of migrating cranes, had been given a photograph of a whooping crane by Ray C. Erickson, head of the whooping crane research program for the Interior Department.

14. Barrows, *Nature's Ghosts*, 322.

15. Ibid., 322–23; Yaffee, *Prohibitive Policy*, 39–42.

16. Barrow, *Nature's Ghosts*, 324 (quoting Carr, "Alligators: Dragons in Distress," 133–38).

17. Barrows, "Dragons in Distress," 269.

18. Ibid., 271 (quoting Hugh Smith, "Report of the Fisheries of the South Atlantic States," *Bulletin of the U.S. Fish Commission* 11 [1893]: 343–45).

19. Barrows, "Dragons in Distress," 271.

20. Ibid., 278. The chief proponent of Florida's legislation was an amateur herpetologist named Ross Allen, who supported his studies by opening the Ross Allen Reptile Institute, an alligator zoo for tourists on the edge of the Ocala National Forest.

21. US Public Law 91-135.

22. Proposed Listing, 35 Fed. Reg. 6069, 6074 (April 14, 1970). Two subspecies of green turtle were proposed for listing: *Chelonia mydas mydas* and *Chelonia mydas japonica*. The three Crocodylia were the black caiman (*Melanosuchus niger*), the Yacare (*Caiman yacare*), and the Morelet's crocodile (*Crocodylus moreletii*).

23. Final Listing, 35 Fed. Reg. 8491, 8495 (June 2, 1970). But four new crocodiles were on the list: the Orinoco crocodile (*Crocodylus intermedius*), the Cuban crocodile (*Crocodylus rhombifer*), the Nile crocodile (*Crocodylus niloticus*), and the gavial (*Gavialis gangeticus*).

24. Final Listing, 35 Fed. Reg. 12222 (July 30, 1970).

25. Monte Lloyd, Robert F. Inger, and F. Wayne King, "On the Diversity of Reptile and Amphibian Species in a Bornean Rain Forest," *American Naturalist* 102, no. 928 (1968): 497–515.

26. Mary Jane Andrei, "The Accidental Conservationist: William T. Hornaday, the Smithsonian Bison Expeditions and the US National Zoo," *Endeavor* 29, no. 3 (2005): 109–13. Hornaday's books include *Two Years in the Jungle*

(1885), *Wild Life Conservation in Theory and Practice* (1914), and the book he dedicated to the US Congress, *Thirty Years War for Wild Life* (1931).

27. F. Wayne King, interview by author, Oct. 14, 2009, Gainesville, FL.

28. F. Wayne King and P. Brazaitis, "Species Identification of Commercial Crocodilian Skins," *Zoologica* 56, no. 2 (1971): 15–70.

29. King interview.

30. At King's suggestion, the commission changed its name to the Species Survival Commission. King interview.

31. Barrow, "Dragons in Distress," 281.

32. Ibid.

33. Ibid. (citing Huey B. Long, F. Wayne King, and Edward R. Ricciuti, eds., *Proceedings of the American Alligator Council, 1968–1969*, n.p.; and Martin Tolchin, "Lindsay Vetoes a Rise in His Pay," *New York Times*, Dec. 30, 1969).

34. F. Wayne King, "Adventures in the Skin Trade," *Natural History* 8 (1971): 10.

35. Ibid., 12.

36. Ibid., 16.

37. A. E. Nettleton Co. v. Diamond, 27 N.Y. 2d 182, 264 N.E.2d 118 (NY 1970).

38. Palladio Inc. v. Diamond, 321 F. Supp. 630 (S.D.N.Y., Nov. 25, 1970), *affirmed per curiam*, 440 F.2d 1319 (U.S. Ct. Appeals, 2d Circuit, March 26, 1971), *cert. denied*, 40 U.S.L.W. 3264 (U.S., Dec. 7, 1971).

39. F. Wayne King, "Historical Review of the Decline of the Green Turtle and the Hawksbill," in *Biology and Conservation of Sea Turtles*, ed. Karen A. Bjorndal, rev. ed. (Washington, DC: Smithsonian Institution Press, 1995), 185.

Chapter 11: You Lost the Turtle Boat

1. Robert M. Ingle, "Florida's Sea Turtle Industry in Relation to Restrictions Imposed in 1971," in *Summary of Florida Commercial Marine Landings, 1971*, Contribution no. 201, Florida Department of Natural Resources, Marine Research Laboratory, Frank Lund private collection. The turtles on board were probably taken from Honduras waters. When the British fishing treaty with Nicaragua ended in the mid-1960s, the *A. M. Adams* and the two other remaining schooners could no longer catch turtles on the Miskito Bank. See Matthiessen, "To the Miskito Bank."

2. Wright Langley, "Turtle Industry Facing Extinction," *Miami Herald*, March 29, 1971, 1B.

3. "Turtle Boat Is Out of Action," *Key West Citizen*, March 26, 1971.

4. Ibid.

5. Langley, "Turtle Industry Facing Extinction."

6. Robert M. Ingle and Frank G. Walton Smith, *Sea Turtles and the Turtle Industry of the West Indies, Florida, and the Gulf of Mexico*, Special Publication, University of Miami, 1949. This book was revised and republished as Thomas P. Rebel, *Sea Turtles and the Turtle Industry of the West Indies, Florida, and the Gulf of Mexico* (Coral Gables: Univ. of Miami Press, 1974).

7. Laws of Florida 1953, chap. 28145, subsec. 12, para. 1.

8. Archie Carr and Robert M. Ingle, "The Green Turtle (*Chelonia mydas mydas*) in Florida," *Bulletin of Marine Science of the Gulf and Caribbean* 9, no. 3 (1959): 315–20. The bulletin is now published as the *Bulletin of Marine Science*.

9. Carr, *Handbook of Turtles*, 356. Carr quotes John James Audubon as mentioning that the Halifax and Indian Rivers hosted large numbers of green turtles in the early 1800s, writing that "great numbers are killed by turtlers and Indians, as well as by cougars, lynxes, bears and wolves," and that one man was alleged to have caught in one year's time 800 turtles by pegging (spearing turtles from a canoe).

10. Carr and Ingle, "Green Turtle in Florida," 318.

11. Carr probably shared this early theory with Wayne King in 1960, when King was a research assistant at Tortuguero. King later argued that sperm storage accounted for the decreased fertility of eggs laid by captive green turtles at Mariculture, Ltd. The sperm storage theory was never proven, however, and it is now known that fertilization takes place in the year that the eggs are laid. David William Owens, telephone interview by author, Sept. 15, 2010.

12. Carr and Ingle, "Green Turtle in Florida," 318.

13. Ibid., 319.

14. Laws of Florida 1957, chap. 57-771. Carr had pointed out this loophole in the paper he wrote with David Caldwell, which Caldwell presented at the Twenty-Second North American Wildlife Conference in Washington, DC, in March 1957. Caldwell and Carr, "Sea Turtle Fishery in Florida," 458.

15. Laws of Florida 1959, chap. 59-483; Fla. Stat. Ann., Section 370.02.

16. See Ross Witham and C. R. Futch, "Early Growth and Oceanic Survival of Pen-Reared Sea Turtles," *Herpetologica* 33 (1977): 404–9. Florida's restocking program ended in 1989 after over 18,000 green turtles, ages six to twelve months, were released. When the program ended, there was not yet any evidence that the restocked green turtles had returned to nest on Florida's beaches. The department decided it had released numbers sufficient to determine if the program had been a success—i.e., if any tagged turtles had returned—and that its priorities were now to reduce adult turtle mortality in shrimp trawling and degradation of nesting and foraging habitats. J. Alan Huff, "Florida (USA) Terminates 'Headstart' Program," *Marine Turtle Newsletter* 46 (1989): 1–2. Huff had worked on the recovery program with Ross Witham and was now the su-

pervisor of the Resource Recovery and Assessment Section of Florida's Bureau of Marine Research.

17. Robert M. Ingle (Florida Department of Natural Resources) to US Department of Interior, May 27, 1970 (emphasis added), in Frank Lund private collection. Ingle sent carbon copies to three members of Congress and to the Southeastern Fisheries Association.

18. J. P. Linduska (associate director, US Fish and Wildlife Service) to Robert M. Ingle, June 5, 1970, Lund collection.

19. US Department of the Interior, Bureau of Sport Fisheries and Wildlife, Rare and Endangered Fish and Wildlife of the United States, Sheet RAP-14: Jan. 1966, Lund collection. The 1966 classification is reprinted in appendix A.

20. Ibid.

21. Linduska to Ingle, June 5, 1970.

22. Jack Rudloe and Anne Rudloe, *Shrimp: The Endless Quest for Pink Gold* (Saddle River, NJ: FT Press, 2010).

23. Jack Rudloe, *Time of the Turtle* (New York: Penguin Books, 1979), 3–12. The shrimp trawlers in Florida caught primarily Atlantic ridley and loggerhead sea turtles in this manner. Rudloe believed green turtles were seldom caught because they could out-swim the nets. In 1956 in *The Windward Road*, Archie Carr described fishermen harpooning Atlantic ridleys in Florida Bay to use as shark bait.

24. The Bureau of Commercial Fisheries was still in the Department of Interior in May 1970. It was moved to the National Oceanic and Atmospheric Administration in July 1970.

25. On Conservation 70's, see Lyman Rogers, "Conservation 70's: A Concept for Environmental Action," *Science Education* 55, no. 1 (1971): 57–60. Bill Lund was actively involved in a campaign to prevent development along the Loxahatchee River. In January 1969, he convinced Governor Claude Kirk and his cabinet to approve the purchase of a large tract of land along the river for conservation. James D. Snyder, *Life and Death on the Loxahatchee* (Jupiter, FL: Pharos Books, 2002), 105. In 1985, the river was designated under the US Wild and Scenic Rivers Act, and Bill Lund was named conservationist of the year by the Florida Audubon Society. Neil Santaniello, "Bill Lund, Builder, River Preservationist," *South Florida Sun Sentinel*, Jan. 15, 1992.

26. Frank Lund, interview by author, Oct. 13, 2009, Gainesville, FL. See also Pat Cullen, "Marine Hobby Brings Protection to Sea Turtles," *Palm Beach Post*, March 18, 1971, D1.

27. Frank Lund, "The Green Turtle in Florida: Summary of Preliminary Data from Jupiter Island, Florida, 1970–1971," Sept. 1, 1971, Lund collection.

28. Ibid.

29. Robert Ingle, "Beware the Reign of Terror," *Floridian*, n.d., Lund collection.

30. Transcript of Florida governor's cabinet meeting, March 16, 1971, Lund collection. See also "Jupiter Student Challenges State on Turtle Conservation," *Palm Beach Times*, March 5, 1971, D2. Frank Lund had appeared at the meeting of cabinet aides on March 4, 1971, and disagreed with Ingle's proposed length standards. Lund said they should be twenty-nine and forty-one inches, respectively. The newspaper quoted him as saying, "A 31-inch limit to my knowledge would protect nothing that's nesting on the east coast."

31. Frank Lund interview.

32. "Jupiter Student Challenges State," D2.

33. Transcript of governor's cabinet meeting, Lund collection, 78–79.

34. Robert L. Shevin to Randolph Hodges, April 19, 1971, Lund collection.

35. Hodges to Shevin, April 23, 1971, Lund collection.

36. Laws of Florida, chap. 71-145, sec. 1, amending Florida Statutes, sec. 370.12(1)(b), as amended by Laws of Florida, chap. 70–357.

Chapter 12: One Man's Opinion

1. Heimann, *Most Offending Soul Alive*, 374.

2. Davis, *Man Who Saved Sea Turtles*, 180. This account of the 1971 meeting in Morges is a reconstruction based on several sources, including Robert I. Standish, "Meeting of the Marine Turtle Specialist Group of the Survival Service Commission, IUCN Headquarters, Morges, Switzerland, 8–10 March 1971," *Biological Conservation* 1971:77–78; Bustard, *Sea Turtles*, 179–97; papers from the meeting archived in the Carr Papers and Harrisson Papers; and telephone interviews with George Hughes and Peter Pritchard, both of whom were in attendance.

3. List of Endangered Foreign Fish and Wildlife, 35 Fed. Reg. 18319 (Dec. 2, 1970).

4. H. Robert Bustard, "Observations on the Flatback Turtle *Chelonia depressa* Garman," *Herpetologica* 25 (1969): 29–34. In 1988, John Hendrickson published a paper redescribing the flatback as a monotypic species in the genus *Natator*. See R. Zangerl, Lupe P. Hendrickson, and John R. Hendrickson, "A Redescription of the Australian Flatback Sea Turtle, *Natator depressus*," *Bishop Museum Bulletins in Zoology* 7 (1988): 1–69. See also Colin J. Limpus, Emma Gyuris, and Jeffrey D. Miller, "Reassessment of the Taxonomic Status of the Sea Turtle Genus *Natator* (McCulloch, 1908), with a Redescription of the Genus and Species," *Transactions of the Royal Society of South Australia* 112 (1988): 1–9.

5. The independent nation of Papua New Guinea was created from this territory in 1975.

6. Bustard, *Sea Turtles*, 176.

7. H. Robert Bustard, *Kay's Turtles* (London: Collins, 1973), 126–28. The book is about Kay's friendship with several adult green turtles.

8. Bustard, *Sea Turtles*, 179–88.

9. Ibid., 188. Bustard described Hendrickson's success in Hawaii in achieving the first captive nesting behavior, which was later replicated by Mariculture in the Cayman Islands.

10. Hendrickson, "Report on Hawaii, March 1969," Harrisson Papers, IUCN Series, box 7.

11. Bustard, "Turtles in Fiji," 317–18.

12. Mariculture, Ltd., was not yet in production, however; the first farm-raised turtles were slaughtered in October 1972. Fosdick and Fosdick, *Last Chance Lost?* 110.

13. Standish, "Meeting of the Marine Turtle Specialist Group."

14. George Hughes, telephone interview by author, May 21, 2011; George R. Hughes, "Loggerheads and Leatherbacks in the Western Indian Ocean," *Indian Ocean Turtle Newsletter* 11 (Jan., 2010): 25.

15. Standish, "Meeting of the Marine Turtle Specialist Group."

16. Bustard, *Sea Turtles*, 204–6. Bustard was familiar with the massive egg collecting program at the leatherback rookery in Trengganu, Malaysia, and took umbrage at Pritchard's use in his 1971 leatherback report of a statement in Carr's 1968 book *So Excellent a Fishe*, that because the egg traffic could be easily curtailed without great effort or expense, the leatherback was the least seriously threatened of all the sea turtles. Bustard wrote, "If the turtle experts themselves cannot agree on the basis of the available data then hope for a wider dissemination of information is indeed grim." Ibid., 205.

17. Fosdick and Fosdick, *Last Chance Lost?* 134. Jim Wood eventually became director of research at Mariculture and general manager at its successor, Cayman Turtle Farm. David Owens, another University of Arizona graduate student of John Hendrickson's, conducted research on reproductive physiology at the turtle farm. Ibid., 136.

18. Langley, "Green Turtle Industry Facing Extinction," *Miami Herald*, March 29, 1971, 1B.

19. Carr to M. Björklund (scientific and administrative coordinator, IUCN Marine Turtle Specialist Group), June 8, 1972, Nicholas Mrosovsky private collection.

20. Archie Carr, "Great Reptiles, Great Enigmas," *Audubon*, March 1972, 24–34.

21. Laws of Florida, chap. 71-145, sec. 1, amending Florida Statutes, sec. 370.12(1)(b). It would be another three years before the Florida legislature adopted comprehensive turtle protection legislation prohibiting all sea turtle fishing and egg hunting, and requiring trawling vessels to return accidentally caught turtles to the water immediately.

22. Robert Gillette, "Endangered Species: Diplomacy Tries Building an Ark," *Science* 179 (Feb. 23, 1973): 777; Barrows, *Nature's Ghosts*, 336–37. The principal drafters were Lee M. Talbot, an officer of the IUCN, and Wolfgang Burhenne, chair of IUCN's Commission on Legislation. Talbot also was one of the principal drafters of the US congressional bills that became the Endangered Species Act of 1973. Barrows, *Nature's Ghosts*, 337, 339.

23. Peter H. Sand, "Commodity or Taboo? International Regulation of Trade in Endangered Species," *Green Globe Yearbook*, 1997, 20.

24. Barrow, *Nature's Ghosts*, 337.

25. Gillette, "Endangered Species: Diplomacy Tries Building an Ark," 777, 780.

26. Robert Gillette, "Endangered Species: Moving Toward a Cease-Fire," *Science*, 179 (March 16, 1973): 1108. In article 1, paragraph (a), CITES defines "species" to include subspecies and geographically separate populations.

27. Gillette, "Endangered Species: Moving Toward a Cease-Fire."

28. Frank Lund, interview by author, Oct. 13, 2009, Gainesville, FL.

29. King, "Historical Review of the Decline of the Green Turtle and the Hawksbill," 186–87. The Japanese delegate did not formally oppose the listing but encouraged the exporting nation of Panama to reverse a previous position and object. Ibid.

30. Gillette, "Endangered Species: Moving Towards Cease-Fire," 1107–8.

31. Ibid., 1108.

32. Marydele Donnelly, *Sea Turtle Mariculture: A Review of Relevant Information for Conservation and Commerce* (Washington, DC: Center for Marine Conservation, 1994), 48; King, "Historical Review," 187.

33. "Green and Loggerhead Turtles Proposed for Foreign Endangered Species List," Department of the Interior News Release, Jan. 4, 1974 (quoted in Yaffee, *Prohibitive Policy*, 115).

34. Notice of Proposed Rulemaking, 38 Fed. Reg. 35486 (Dec. 28, 1973). The loggerhead was also proposed for listing.

35. Article 7, Convention on International Trade in Endangered Species of Flora and Fauna (CITES), *International Legal Materials*, 12 (1973): 1085 (US Treaty Series, 993:243, signed March 3, 1973). Article 7, paragraph 4, provides that specimens from Appendix I species that were bred in captivity for commercial purposes can be treated as if they were on Appendix II, meaning the ex-

porting country may issue a permit if its scientific and management agencies determine that it is not detrimental to the survival of that species nor taken in violation of its conservation laws.

Chapter 13: Down on the Farm

1. Archie Carr to L. A. Greenwalt (director, Bureau of Sports Fisheries and Wildlife, US Dept. of the Interior), Jan. 22, 1974, attached as exhibit 2 to F. Wayne King affidavit, Harrisson Papers, IUCN Series, box 8, Mariculture Ltd. v. Biggane folder.

2. Alan Sterling Parkes, *Off-Beat Biologist: The Autobiography of Alan S. Parkes* (Cambridge, UK: Galton Foundation, 1985), 285, 287.

3. Ibid., 291. Until May 1973, workers for Mariculture, Ltd., were successfully hatching eggs taken from the nesting beaches in Central America but were not yet able to produce eggs on the farm. The company completed its first slaughter of market-size turtles (80–100 lbs.) in the autumn of 1972 and sold the belly plate to soup makers and the meat and leather to other manufacturers. Ibid., 286.

4. Archie Carr to Polly Bergen, July 29, 1970, Carr Papers, Series 2, box 16.

5. Archie Carr to Antony Fisher, July 10, 1973, reprinted in Fosdick and Fosdick, *Last Chance Lost?* 139–41.

6. J. R. Hendrickson to Edwin Z'berg (chair, California Assembly Natural Resources Committee), Aug. 31, 1973, quoted in Fosdick and Fosdick, *Last Chance Lost?* 144–45.

7. Archie Carr to Roderic Tuttle (consultant, California Assembly Natural Resources Committee), Sept. 12, 1973, reprinted in Fosdick and Fosdick, *Last Chance Lost?* 145.

8. Fosdick and Fosdick, *Last Chance Lost?* 158–59. Ehrenfeld presented a paper entitled "The Food Potential of the Sea Turtle: Can Mariculture Work?" on March 12, 1973, at the New York Academy of Sciences, with Antony Fisher and Henry Hamlin in attendance. Ehrenfeld's paper was published the following year as "Conserving the Edible Sea Turtle: Can Mariculture Help?" *American Scientist* 62 (1974): 23–31.

9. Archie Carr and A. R. Main, "Turtle Farming Project in Northern Australia: Report on an Inquiry into Ecological Implications of a Turtle Farming Project," Department of the Special Ministry of State, Canberra, Oct. 1973. A companion report on the organization and management structure of the turtle farming project was prepared by L. P. Smart, a chartered accountant from Melbourne, Australia. It is appended to the report by Carr and Main.

10. Ibid., 2, para. 1.7, app. B.

11. Ibid., 9.

12. Ibid., 18.

13. Ibid., 18–19.

14. Ibid., v–vi.

15. Fosdick and Fosdick, *Last Chance Lost?* 128–29.

16. Ibid., 129.

17. CITES, art. 14. This discussion between Mariculture's owners and its lawyers is a reconstruction based on Fosdick and Fosdick, *Last Chance Lost?* 127–32 and letters found in the Carr Papers and the Harrisson Papers. The book *Last Chance Lost?* was published privately by Irvin Naylor, the former president of Mariculture and contains a number of his recollections of events, which correspond with the events documented in correspondence found in the Carr and Harrisson Papers.

18. 16 U.S.C. sec. 1533(d).

19. 16 U.S.C. sec. 1533(a)(2). This section applied to "any species over which program responsibilities have been vested in the Secretary of Commerce pursuant to Reorganization Plan Numbered 4 of 1970, 5 U.S.C. §903 note."

20. Robert Schoning to J. P. Gilchrist (California Seafood Institute), Feb. 21, 1974, Harrisson Papers, IUCN Series, box 8.

Chapter 14: Conservation through Commerce

1. F. Wayne King (New York Zoological Society) to Rogers C. B. Morton (Secretary of the Interior), April 23, 1974, attached as exhibit 3 to Affidavit of F. Wayne King in Mariculture Ltd. v. Biggane, in Harrisson Papers, IUCN Series, box 8.

2. Irvin S. Naylor (president, Mariculture), Affidavit filed in Mariculture v. Biggane, Oct. 8, 1974, Harrisson Papers, IUCN Series, box 8.

3. Mariculture Ltd. v. Biggane, 48 A.D.2d 295, 369 N.Y.S.2d 219 (N.Y.S. Ct., App. Div., 1975). Background on the Mason Act, its relationship to the federal ESA, and the history of legal challenges is found in the historical and statutory notes to McKinney's Consolidated Laws of New York Annotated, ECL sec. 11-0536 (2009).

4. A. E. Nettleton Co. v. Diamond, 264 N.E.2d 118 (1970), appeal dismissed *sub nom.* Reptile Products Ass'n v. Diamond, 401 U.S. 969; and Palladio, Inc. v. Diamond, 321 F.Supp. 630 (S.D.N.Y.1970), affirmed 440 F.2d 1319, certiorari denied 404 U.S. 983.

5. See Arthur F. McEvoy, *The Fisherman's Problem: Ecology and Law in the California Fisheries, 1850–1980* (New York: Cambridge Univ. Press, 1986).

6. This account of the enactment of California's turtle mariculture bill is from R. M. Christensen, "Special Report: Green Sea Turtle Farming," *Chelo-*

nia 3, no. 2 (Aug. 1976): 3–6, and letters between A. J. Mence (IUCN Survival Service Commission) and Tom Harrisson as co-chair of the Marine Turtle Specialist Group, Harrisson Papers, IUCN Series, box 8.

7. Fosdick and Fosdick, *Last Chance Lost?* 170.

8. US Fish and Wildlife Service and National Marine Fisheries Service, Sea Turtles, Proposed Rule to List Green, Loggerhead, and Pacific Ridley Sea Turtles as Threatened Species under Similarity of Appearance Provision, 41 Fed. Reg. 24378 (June 16, 1976).

9. The ESA of 1973, sec. 4(e), allows the secretaries of interior and commerce, by regulation, to treat any unlisted species as endangered or threatened if enforcement personnel would have substantial difficulty in differentiating between listed and unlisted species. 16 U.S.C. sec. 1533(e); Code of Federal Regulations, Title 50, sec. 17.50.

10. Fosdick and Fosdick, *Last Chance Lost?* 130–31.

11. US Fish and Wildlife Service and National Marine Fisheries Service, Sea Turtles, Notice of Status Review and Request for Information, 39 Fed. Reg. 29605, 29607 (Aug. 16, 1974).

12. Robert W. Schoning (director, NMFS) to John Gilchrist (California Seafood Institute), Feb. 21, 1974, Harrisson Papers, IUCN Series, box 8, Correspondence, Jan.–June 1974 folder.

13. The ESA of 1973 also provides an exception for captive-bred, self-sustaining populations held not for commercial purposes. 16 U.S.C. sec. 1538(b) (1). The Fish and Wildlife Service published regulations to implement this in Title 50, Code of Federal Regulations sec. 17.7 (deleted in 1979). See Stephen M. Fernandez, "Captive-Bred Exceptions: An Unconventional Approach to Conservation under the Endangered Species Act," *University of Florida Journal of Law & Public Policy* 15 (2003): 155–93.

14. The August 1974 memorandum broadly defined the agencies' respective roles but left unresolved the allocation of responsibilities over sea turtles. They did not conclude an understanding on sea turtles until July 1977. Yaffee, *Prohibitive Policy*, 115, 212.

15. Fosdick and Fosdick, *Last Chance Lost?* 131. In 1976, NMFS was embroiled in a very controversial permit decision concerning the accidental drowning of dolphins by American tuna fishing vessels in the eastern tropical Pacific. The agency was required to set a quota on the number of dolphins that could be killed without depleting the population below its optimum sustainable population as defined by the Marine Mammal Protection Act of 1972. NMFS had almost no data on which to base this number and was under pressure to keep the tuna fleet fishing. To avoid listing the dolphin species as threatened under ESA, NMFS classified them as depleted under the Marine Mammal Protection Act

and issued a permit to give the industry time to develop a technical solution to the dolphin by-catch. This avenue was not available for the sea turtle by-catch problem because marine reptiles are not subject to the Marine Mammal Protection Act. But NMFS needed to buy time while its research and technical staff sought solutions to fishery conflicts with marine wildlife—specifically, devices to exclude turtles from shrimp trawling nets. The threatened classification would give the agency latitude to take its technical and managerial approach to wildlife. Yaffee, *Prohibitive Policy*, 114, citing an interview with a former director of the NMFS office responsible for the endangered species program.

16. The following discussion of the plans for the IUCN Principles and Recommendations is based on several letters and memoranda written by Tom Harrisson and Tony Mence, executive secretary of the IUCN's Survival Service Commission in 1974. Harrisson Papers, IUCN Series, box 8.

17. The IUCN Principles and Recommendations are reprinted in appendix B.

18. Parkes, *Off-Beat Biologist*, 291. Parkes describes how two short papers he had written after the successful captive mating were combined into a "paper" without his knowledge and published in the Cayman Islands' monthly magazine, *Northwester*, to convince the local banks and other investors that the farm was a worthwhile investment. Ibid., 292. Parkes gave a similar account in letters to Tom Harrisson in the months following the ad hoc task force's meeting at Mariculture in November 1974. Harrisson Papers, IUCN Series, box 8.

19. Parkes, *Off-Beat Biologist*, 297.

20. The following discussion of the task force visit to Mariculture is drawn from ibid., 298–99.

21. Anthony J. Mence, Draft Report to the Chairman of the Survival Service Commission by the Task Force on the Commercial Exploitation of Marine Turtles, Annex I, Dec. 17, 1974, Harrisson Papers, IUCN Series, box 8. Quotations in the text are from this report.

22. King's letter to Fullerton is described in a letter Leo Brongersma sent to King taking him to task for sending it without consulting other members of the task force or the Survival Service Commission. Leo Brongersma to F. W. King, Jan. 30, 1975, Harrisson Papers, IUCN Series, box 8.

23. Ibid.

24. Henry Reichart to State Senator Behr, Dec. 1974, Harrisson Papers, IUCN Series, box 8.

25. F. G. Nicholls (IUCN) to E. C. Fullerton, May 15, 1975, cited in J. J. Zedrosser (New York assistant attorney general) to US Fish and Wildlife Service, July 13, 1975, commenting on the proposed threatened status of sea turtles, Harrisson Papers, IUCN Series, box 8.

26. Fosdick and Fosdick, *Last Chance Lost?* 187, 191.

27. Ibid., 166, 178, 188.

28. US Fish and Wildlife Service and National Marine Fisheries Service, Sea Turtles, Proposed "Threatened" Status, 40 Fed. Reg. 21974 (May 20, 1975). The two services also proposed to list the loggerhead and Pacific ridley turtles as threatened.

29. The following account of the conversation between Jones and Moyle is a reconstruction based on Fosdick and Fosdick, *Last Chance Lost?* 189–91.

30. 40 Fed. Reg. 21983 (May 20, 1975). The other exceptions to the ban on sale, possession, or transport of these species were for incidental catch during fishing or research activities, provided the fishing was not directed towards sea turtles or conducted in an "area of substantial breeding or feeding." Turtles were required to be returned to the water immediately, whether dead or alive, with due care to minimize injuries to live specimens. There was no exception for subsistence food purposes. Ibid., 21983–84.

31. To obtain a mariculture permit, the company would have to mark its products in such a way that the marking would stay with the product until after the retail sale or export. The country of origin would also have to certify that it had laws protecting the green turtles and that the removal of eggs or turtles from it "will not be detrimental to the survival of the species in the wild." Ibid., 21985.

32. The motto "Conservation through Commerce" appeared prominently on the letterhead used by Mariculture's president, Irvin Naylor. Some members of the Marine Turtle Specialist Group must have taken particular umbrage to the phrase.

33. Fosdick and Fosdick, *Last Chance Lost?* 192.

34. F. G. Nicholls to E. C. Fullerton, Sept. 25, 1975, quoted in Christensen, "Special Report," 4.

35. Ibid. The permits issued to Mariculture and six wholesalers were retroactive to Sept. 25, 1975, with an expiration date of Dec. 31, 1976.

36. E. C. Fullerton to NMFS and FWS, July 7, 1975, quoted in Christensen, "Special Report," 4.

37. Fosdick and Fosdick, *Last Chance Lost?* 202–3; Christensen, "Special Report," 2.

38. Minutes of Dusseldorf meeting in F. Wayne King to Peter Scott, Sept. 1977, Carr Papers, Series 3, box 28.

Chapter 15: The Best Available Science

1. Yaffee, *Prohibitive Policy*, 114–17. Congress was aware of the different organizational goals of the two agencies and attempted to allocate responsibilities

accordingly in the Endangered Species Act of 1973. Section 4(a)(2) as enacted provided that for species "over which program responsibilities have been vested in the Secretary of Commerce pursuant to Reorganization Plan Numbered 4 of 1970," in cases where the secretary of commerce determines that a species should be listed as endangered or threatened or changed from threatened to endangered, he must inform the secretary of interior, "who shall list such species in accordance with this section." If the secretary of commerce believes a species should be removed from either list or downlisted from endangered to threatened, he is to make a recommendation to that effect to the secretary of interior, who is to implement the recommendation *if he concurs.* In no event may the secretary of interior list, remove from any list, or change the status of any species for which the secretary of commerce has programmatic responsibility without a favorable determination by the secretary of commerce. Sec. 4(a)(2), Pub. L. 93-205, 87 Stat. 884 (Dec. 28, 1973).

2. Lynn Greenwalt (FWS director) to Robert Schoning (NMFS director), Aug. 18, 1975, quoted in Yaffee, *Prohibitive Policy,* 115–16.

3. Yaffee, *Prohibitive Policy,* 116.

4. This reconstruction of the hearing on the draft EIS on the proposed listing is based on Yaffee, *Prohibitive Policy,* 114–17, and the draft and final EISs prepared by NMFS. US Department of Commerce, *Draft Environmental Impact Statement: Proposed Listing of the Green Sea Turtle* (Chelonia mydas), *Loggerhead Sea Turtle* (Caretta caretta), *and Pacific Ridley Sea Turtle* (Lepidochelys olivacea) *as Threatened Species under the Endangered Species Act of 1973,* National Marine Fisheries Service, Feb. 11, 1976; US Department of Commerce, *Final Environmental Impact Statement: Listing and Protecting the Green Sea Turtle* (Chelonia mydas), *Loggerhead Sea Turtle* (Caretta caretta), *and Pacific Ridley Sea Turtle* (Lepidochelys olivacea) *Under the Endangered Species Act of 1973,* National Marine Fisheries Service, July 1978.

5. Yaffee, *Prohibitive Policy.*

6. This summary of King's testimony is based on the written comments he submitted on the draft EIS on March 10, 1976, which are summarized in NMFS's *Final Environmental Impact Statement* (July 1978), 116–17, and in Fosdick and Fosdick, *Last Chance Lost?* 217–19.

7. William A. Butler, "Statement of the Environmental Defense Fund on the Proposed Listing of the Green, Loggerhead, and Pacific Ridley Sea Turtles as Threatened Species," Feb. 25, 1975, 4, 8, Carr Papers, Series 4, box 38, folder 5.

8. Ibid, 8.

9. Fosdick and Fosdick, *Last Chance Lost?* 219–21.

10. Michael C. Lipske, "Sea Turtles Suffer as Bureaucrats Bicker," *Defenders* 52 (Aug. 1977): 228, cited in Yaffee, *Prohibitive Policy,* 116, 213.

11. US Fish and Wildlife Service and National Marine Fisheries Service, "Sea Turtles: Proposed Regulations Treating Three Species as Threatened under the 'Similarity of Appearance' Clause," 41 Fed. Reg. 24378-82 (June 16, 1976).

12. The Swiss had proposed moving the Pacific hawksbill to Appendix I, and the Australians proposed to uplist the leatherback. At the plenipotentiaries' meeting in 1973, the parties agreed only that international trade was threatening the survival of the Atlantic hawksbill, *Eretmochelys imbricata imbricata*, and the Atlantic ridley, *Lepidochelys kempii* (now known as Kemp's ridley). The leatherback, *Dermochelys coriacea*, and the other genera of *Cheloniidae*—the loggerhead, *Caretta caretta*; the Pacific hawksbill, *Eretmochelys imbricata squamata*; the Pacific ridley, *Lepidochelys olivacea*; and the green turtle, *Chelonia mydas*, with the exception of Australian population—were on Appendix II. The 1976 UK proposal suggests that the United Kingdom believed green turtle products could be shipped from the Cayman Islands because they met the "bred in captivity" exception to Appendix I found in article VII, paragraph 4.

13. Robert E. Stevens (NMFS endangered species administrator) to H. M. Hutchings (acting assistant director for fisheries management, NMFS), Nov. 12, 1976, Carr Papers, Series 3, box 38, folder 1.

14. Ibid., 6. Stevens reported that even if the United States vote had been no, the proposal would have had 70 percent of the votes in the affirmative.

15. When France and Italy formally became parties to CITES, in 1978 and 1979, respectively, they also filed reservations on the Appendix I listing of green turtles. Daniel Navid, "Conservation and Management of Sea Turtles: A Legal Overview," in Bjorndal, *Sea Turtles*, 525; F. Wayne King, "Historical Review of the Decline of the Green Turtle and the Hawksbill," 187.

16. Stevens to Hutchings, Nov. 12, 1976, 6–7. NMFS made the decision not to change the proposed ESA listing but to rewrite the mariculture permit regulations to conform to the CITES requirement of a certificate from the exporting country. Prudence Fox (NOAA Office of International Affairs), Memorandum to the Record, Nov. 9, 1976, Carr Papers, Series 3, box 38, folder 1.

17. The EPA had reviewed the EIS and rated it as inadequate, requiring NMFS to take special care to address the perceived deficiencies in order not to encourage a National Environmental Policy Act lawsuit by the disappointed parties.

18. US Fish and Wildlife Service, "CITES Implementing Regulations," 42 Fed. Reg. 10462-88 (May 23, 1977). The regulations prohibited commercial importation of green turtles and their products, with exemptions for animals bred in captivity, exported from Australia, or qualifying under the CITES "grandfather" clause (i.e., acquired pre-Convention). NMFS, *Final Environmental Impact Statement*, 84, 92.

19. Memorandum, July 18, 1977, Carr Papers, Series 3, box 38, folder 1.

20. Yaffee, *Prohibitive Policy*, 213, citing *US Senate Committee on Environment and Public Works, 1977 Endangered Species Act Oversight Hearings*, serial no. 95-H33, 95th Cong., 1st sess., 66.

21. On February 28, 1978, the Environmental Defense Fund asked NMFS to reopen the public comment period so it could submit information it had acquired in the nearly three years since the publication of the proposed listing regulations. Much of the new information concerned the progress on the farm toward a self-sufficient breeding herd. The agencies granted the request on March 27, 1978, but denied other parties' requests to extend the comment period again. US Fish and Wildlife Service and National Marine Fisheries Service, "Final Rules: Listing and Protecting Loggerhead Sea Turtles as 'Threatened Species' and Populations of Green and Olive Ridley Sea Turtles as Threatened Species or 'Endangered Species,'" 43 Fed. Reg. 32801 (July 28, 1978).

22. The turtle farm was shipping products to markets in Germany because Germany had filed a reservation to the CITES decision to list the green turtle on Appendix I. The United Kingdom, of which the Cayman Islands was an overseas territory or dependency, certified the products as being from turtles that were "bred in captivity."

23. NMFS, *Final Environmental Impact Statement*, 101. Of the forty-four parties who had commented on the proposed mariculture exception, twenty-four opposed and twenty supported the exception. FWS and NMFS, "Final Rules: Listing and Protecting Sea Turtles," 32804.

24. NMFS, *Final Environmental Impact Statement*, 117–19.

25. The letters submitted are summarized in Fosdick and Fosdick, *Last Chance Lost?* 239–46.

26. Farm CORAIL (Compagnie Reunionnaise d'Aquaculture et d'Industries Littorales) was at St. Leu on the island of Réunion, an overseas department of France in the southwestern Indian Ocean. It began as an experimental breeding facility in 1972. Farm CORAIL took daylight-emerging hatchlings from nature reserves on the islands of Tromlein and Europa for captive rearing in the facility at St. Leu. CORAIL sent the meat and calipee to France for soup production, and shops on Réunion sold polished shells to tourists and for jewelry-making. Donnelly, *Sea Turtle Mariculture*, 62–63; Mrosovsky, *Conserving Sea Turtles*, 148. The French government argued that CORAIL products could be exported to other CITES parties such as the United Kingdom and West Germany under the "bred in captivity" exception to Appendix I. Australia still had hope that the green turtle farms on the Torres Strait Islands would succeed. If the CITES parties approved downlisting the source populations on Tromlein and Europa Islands for Farm CORAIL to Appendix II, it would set a precedent which Aus-

tralia could later use when the Torres Strait turtle products were ready for market.

27. Alan Gill, "Enter the Political Turtle," *ABM Review* 64, no. 3 (1974): 22–24; Margaret Jones, "Turtle Farming: Good Intentions but Disappointing Results," *ABM Review* 64, no. 3 (1974): 18–21; Carr and Main, "Turtle Farming in Northern Australia."

28. It was not until August 1979 that the minister for aboriginal affairs announced that the farms were to be phased out, after learning from Applied Ecology Proprietary, Ltd., that the project was not likely to become commercially viable. Mrosovsky, *Conserving Sea Turtles*, 39 (citing Aug. 23, 1979, press release from Senator F. M. Chaney, minister for aboriginal affairs); Daley, *Changes in Great Barrier Reef*, 428.

29. Vermont Yankee Nuclear Power Corp. v. Natural Resources Defense Council, Inc., 435 U.S. 519, 545–46 (1978), cited in Cayman Turtle Farm, Ltd., v. Andrus, 478 Fed. Supp. 125, 131 (D.D.C., 1979).

30. US Fish and Wildlife Service and National Marine Fisheries Service, Decision Memorandum, Dec. 5, 1978, quoted in Fosdick and Fosdick, *Last Chance Lost?* 248–52.

31. Henry A. Reichart, "Farming and Ranching as a Strategy for Sea Turtle Conservation," in Bjorndal, *Sea Turtles*, 467.

32. Conf. Res. 2.12, CITES, Second Conference of the Parties, San José, Costa Rica, March 19–30, 1979. The French delegation introduced amendments only to clarify the grammar in the French text.

33. Summary Report of the Plenary Session, Second Meeting of the Conference of the Parties, March 19–30, 1979, Plen. 2.12, reprinted in *Hearing before the Subcommittee on Fisheries and Wildlife Conservation and the Environment, Committee on Merchant Marine and Fisheries, House of Representatives on United States Policy Relating to Ranching, Farming, Mariculture, and Aquaculture of Species of Fish and Wildlife and the Endangered Species Act,* Oct. 4, 1982, serial no. 97-44 (Washington, DC: Government Printing Office, 1982), 71–75.

34. Schulz and Reichart described the data from Schulz's 1975 paper showing that green turtles lay approximately one million eggs each year on Suriname's beaches, and that one-quarter of them are destroyed by spring tides if the nests are left unattended. See Joop P. Schulz, "Sea Turtles Nesting in Suriname," *Stichting Natuurbehoud Surinam (STINASU) Verhandeling* 3 (1975): 1–143.

35. Reichart, "Farming and Ranching as a Strategy," 469–70.

36. Ibid., 467.

37. Cayman Turtle Farm, Ltd., v. Andrus, 478 Fed. Supp. 125 (D.D.C. 1979).

38. William A. Johnson, "Cayman Turtle Farm Ltd.: The Crock of Gold," *British Herpetological Bulletin* 2 (1980): 20–22.

39. The Supreme Court had recently ruled in *TVA v. Hill*, 437 U.S. 153 (1978), the infamous snail darter case, that the secretaries were supposed to give endangered species the benefit of the doubt in technical questions arising under the ESA—for example, whether closing a huge hydropower dam would wipe out the tiny snail darter fish. However, Congress had amended the act to provide a means for exempting certain projects from the act's prohibitions.

Chapter 16: A Global Strategy

1. Carr, "Notes for an Action Plan for Marine Turtles," Carr Papers, Series 4, box 45, folder 13.

2. Henk Reichart, telephone interview by author, June 28, 2011.

3. David W. Owens, telephone interview by author, Sept. 15, 2010. Leo Brongersma had sent John Hendrickson a copy of a negative review that someone had proudly sent him, not realizing that Brongersma was a supporter of Cayman Turtle Farm and a friend of Hendrickson's and his students.

4. Ibid. Nicholas Mrosovsky, by now editor of the *Marine Turtle Newsletter*, would later criticize Florida's program as being driven more by sentimentality and publicity-seeking than science. See Mrosovsky, *Conserving Sea Turtles*, 33.

5. Archie Carr, "Tom Harrisson: Obituary," IUCN/SSC *Marine Turtle Newsletter* 1 (1976): 4–5.

6. Mrosovsky, *Conserving Sea Turtles*, 44–50. Mrosovsky used the hatchlings he obtained from the South Carolina Department of Natural Resources and from Georgia to examine the seasonal sex ratio on a natural beach. Nicholas Mrosovsky, Sally R. Hopkins-Murphy, and James I. Richardson, "Sex Ratios of Sea Turtles: Seasonal Changes," *Science* 225 (Aug. 17, 1984): 739–41. He also got eggs from Joop Schulz's project in Suriname. C. L. Yntema and Nicholas Mrosovsky, "Incubation Temperature and Sex Ratio in Hatchling Loggerhead Turtles: A Preliminary Report," *Marine Turtle Newsletter* 11 (1979): 9–10.

7. David W. Owens, "Introduction to the Symposium," *American Zoologist* 20, no. 3, Behavioral and Reproductive Biology of Sea Turtles (1980): 485–86; Owens, telephone interview, Sept. 15, 2010.

8. Jeremy Cherfas, "The Song of the Turtle," *New Scientist* 84 (Dec. 13, 1979): 880–82.

9. David Ehrenfeld, "Options and Limitations in the Conservation of Sea Turtles," in Bjorndal, *Sea Turtles*, 457 (also quoted in Karen Bjorndal's "Introduction to the First Edition," *Sea Turtles*, 7). Bjorndal's *Sea Turtles* was the proceedings of the conference.

10. George H. Balazs, "Growth Rates in Immature Green Turtles in the Hawaiian Archipelago," in Bjorndal, *Sea Turtles*, 122.

11. Bjorndal, "Introduction," 7–8.

12. George R. Hughes, "Nesting Cycles in Sea Turtles—Typical or Atypical?" in Bjorndal, *Sea Turtles*, 81–89.

13. King, "Historical Review of the Decline of the Green Turtle," 184–85. Curiously, King makes no reference to James Parsons' 1962 book, *The Green Turtle and Man*, which documented the extent of green turtle exploitation.

14. Bernard Nietschmann, "The Cultural Context of Sea Turtle Subsistence Hunting in the Caribbean and Problems Caused by Commercial Exploitation," in Bjorndal, *Sea Turtles*, 443. See also Nietschmann, *The Caribbean Edge*.

15. Judith Fitzpatrick (formerly Nietschmann), telephone interview by author, Feb. 9, 2010.

16. See D. R. Gross, "Review of *The Turtle People*, a film by Brian Weiss," *American Anthropologist* 76 (1974): 486–87.

17. When war did come to Nicaragua, Nietschmann became an advocate for regional autonomy for the Miskito. See Bernard Nietschmann, *The Unknown War: The Miskito Nation, Nicaragua and the United States* (New York: Freedom House and Univ. Press of America, 1989).

18. Nietschmann, "Sea Turtle Subsistence Hunting," 444. Nietschmann told the audience that once the factories closed, the Miskito turned back to agriculture and subsistence hunting and fishing. This helped them to survive when the revolution broke out and the shipment of food and supplies from eastern Nicaragua was disrupted.

19. Ehrenfeld, "Options and Limitations in the Conservation of Sea Turtles," 458.

20. For example, Ehrenfeld, "Conserving the Edible Sea Turtle," 23–31.

21. Ehrenfeld, "Options and Limitations," 461.

22. Reichart, "Farming and Ranching as a Strategy," 467.

23. Nicholas Mrosovsky, interview by author, Aug. 12, 2010, Toronto. See also Mrosovsky's editorial, *Marine Turtle Newsletter* 13 (Nov. 1979): 1–4.

24. Mrosovsky, Editorial, 1.

25. Mrosovsky, interview by author; F. Wayne King, interview by author, and e-mail to author, June 23, 2011. King said he had cancelled Mrosovsky's appointment because under IUCN's governance policies, Tony Mence, the Survival Service Commission's executive officer, did not have the authority to appoint the chair of a species specialist group. After King's action, George Balazs was appointed vice-chair of the Marine Turtle Specialist Group.

26. C. Kenneth Dodd, Jr., "Does Sea Turtle Aquaculture Benefit Conservation?" in Bjorndal, *Sea Turtles*, 479. Dodd's reference to "optimum yield" had

to do with the attempt by some farms to set a quota on the number of doomed eggs to be taken for the farm or ranch. Dodd cited the 1975 paper by Joop Schulz in which Schulz acknowledged that the egg quota for the Suriname farm was necessarily arbitrary given the unknowns regarding the age distribution, life span, and total reproduction of the wild population. Schulz wrote that "even if we could establish how many hatchlings are required to produce one mature female and how many eggs one female produces during her lifetime, there are no quantitative data available on mortality (including the catches in Brazilian waters)." Schulz, "Sea Turtles Nesting in Suriname."

27. Dodd, "Does Sea Turtle Aquaculture Benefit Conservation?" 478 (citing the farm's comments on reconsideration of the July 1978 final rules and Dodd's memorandum to FWS in response to these comments.)

28. David Mack, Nicole Duplaix, and Susan Wells, "Sea Turtles, Animals of Divisible Parts: International Trade in Sea Turtle Parts," in Bjorndal, *Sea Turtles*, 545–62.

29. Jeremy Cherfas, "No End to Trade in Turtles," *New Scientist* 84 (Dec. 13, 1979): 852; John Pickett and Simon Townson, "Political Problems for the Cayman Turtle Farm: Which Way Conservation?" *British Herpetological Society Bulletin*, no. 1 (1980): 18–20.

30. Information on Judith Mittag's experience at the 1979 conference is from Fosdick and Fosdick, *Last Chance Lost?* 266.

31. Cherfas, "No End to Trade in Turtles."

32. Fosdick and Fosdick, *Last Chance Lost?* 266.

33. Nicholas Mrosovsky, "MTN 100: Looking Back, Looking Forward," *Marine Turtle Newsletter* 100 (2003): 3–4.

34. Ibid., 3.

35. Many years later, Peter Pritchard, who served as executive officer of the Marine Turtle Specialist Group in its early years and was Carr's friend and colleague for many years reported that Carr had in fact been less than friendly to former associates who condoned the killing of sea turtles. See Virginia Smith, "Under the Shell: Review of *Tales from the Thébaïde: Reflections of a Turtle-man* by Peter C. H. Pritchard," *Times Literary Supplement*, Nov. 9, 2007. In an interview for the review of his 2007 book, Pritchard acknowledged that Carr "was quite frank about his emotional attachment to his creatures when questioned by a newspaper reporter a month before he died in 1987: 'I just like the looks of their faces,' he replied." Pritchard believes that as Carr grew older, his emotional attachment caused him to give up eating turtle meat and oppose turtle farming. Carr simply "could not abide their killing for any reason, and broke off relationships with those who felt otherwise."

36. The symposium papers were published in a special issue of the *American Zoologist* 20, no. 3, Behavioral and Reproductive Biology of Sea Turtles (1980): Archie Carr, "Some Problems of Sea Turtle Ecology," 489–98; Peter C. H. Pritchard, "The Conservation of Sea Turtles: Practice and Problems," 609–17; Harold F. Hirth, "Some Aspects of the Nesting Behavior and Reproductive Biology of Sea Turtles," 507–23; Ross Witham, "The 'Lost Year' Question in Young Sea Turtles," 525–30; John R. Hendrickson, "The Ecological Strategies of Sea Turtles," 597–608.

37. Nicholas Mrosovsky, "Dedication to Archie Carr," *American Zoologist* 20, no. 3 (1980): 487–88.

38. Nicholas Mrosovsky, "Thermal Biology of Sea Turtles," *American Zoologist* 20, no. 3 (1980): 531–47.

39. Mrosovsky, "Dedication," 488.

40. Papers based on research conducted at Cayman Turtle Farm that were presented at the ASZ symposium included David W. Owens, "The Comparative Reproductive Physiology of Sea Turtles," *American Zoologist* 20, no. 3 (1980): 549–63; and James R. Wood and Fern E. Wood, "Reproductive Biology of Captive Green Sea Turtles *Chelonia mydas*," ibid., 499–505.

41. Cayman Turtle Farm, Ltd., v. Cecil D. Andrus, Secretary of the Department of Interior, 478 Fed. Supp. 125 (D.D.C. 1979), affirmed *per curiam* by an unpublished opinion (D.C. Cir. 1980).

42. George Mackinnon Papers, box 124, Cayman Turtle Farm v. Sec'y of Interior, No. 79-2031, Minnesota Historical Society Library, St. Paul, MN.

Epilogue: Supply and Demand

1. Clem Tisdell, "Conflicts about Living Marine Resources in Southwest Asian and Australian Waters: Turtles and Dugongs as Cases," *Marine Resource Economics* 3, no. 1 (1986): 89–109.

Bibliography

Aldrich, James L., and Anne M. Blackburn. "A Tribute to Harold J. Coolidge." *Environmentalist* 5, no. 2 (1985): 83–84.

Andrei, Mary Anne. "The Accidental Conservationist: William T. Hornaday, the Smithsonian Bison Expeditions, and the US National Zoo." *Endeavor* 29, no. 3 (2005): 109–13.

Balazs, George H. "Growth Rates in Immature Green Turtles in the Hawaiian Archipelago." In Bjorndal, *Sea Turtles*, 117–25.

Banks, Edward. "The Breeding of the Edible Turtle (*Chelonia mydas*)." *Sarawak Museum Journal* 44, no. 4 (1937): 523–32.

———. *A Naturalist in Sarawak*. Kuching, Sarawak: Kuching Press, 1949.

Barrett, Charles. "The Great Barrier Reef and Its Isles: The Wonder and Mystery of Australia's World-Famous Geographical Feature." *National Geographic* 58, no. 3 (1930): 354–84.

Barrows, Mark V., Jr. "Dragons in Distress: Naturalists as Bioactivists in the Campaign to Save the American Alligator." *Journal of the History of Biology* 42, no. 2 (2009): 267–88.

———. *Nature's Ghosts: Confronting Extinction from the Age of Jefferson to the Age of Ecology*. Chicago: University of Chicago Press, 2009.

Bennett, Peter, and Ursula Keuper-Bennett. *The Book of Honu: Enjoying and Learning about Hawaii's Sea Turtles*. Honolulu: University of Hawaii Press, 2008.

Benson, Etienne. "A Difficult Time with the Permit Process." *Journal of the History of Biology* 44, no. 1 (2011): 103–23.

Bjorndal, Karen A. ed., *Biology and Conservation of Sea Turtles*. Proceedings of the World Conference on Sea Turtle Conservation, Washington, D.C., 26–30 November 1979. Rev. ed. Washington, DC: Smithsonian Institution Press, 1995.

———. "Introduction to the First Edition." In Bjorndal, *Sea Turtles*, 7–8.

Bodden, Joyann. "Captain Allie's Memories: Looking Back on 55 Years at Sea," *Northwester*, May 1973, 51–52.

Brice, John J. "The Fish and Fisheries of the Coastal Waters of Florida." In *Report of the U.S. Commissioner of Fish and Fisheries for the Year Ending June 30, 1896*, app. 6. Washington, DC: Government Printing Office, 1897.

Broderick, Annette C., et al. "Are Green Turtles Globally Endangered?" *Global Ecology and Biogeography* 15 (2006): 21–26.

Burhenne, Wolfgang E. "The Draft Convention on the Import, Export, and Transit of Certain Species." *Biological Conservation* 1968: 61–62.

Bustard, H. Robert. *Kay's Turtles*. London: Collins, 1973.

———. "Observations on the Flatback Turtle *Chelonia depressa* Garman." *Herpetologica* 25 (1969): 29–34.

———. *Sea Turtles: Natural History and Conservation*. London: Collins, 1972.

———. "Should Sea Turtles Be Exploited?" *Marine Turtle Newsletter* 15 (1980): 3–5.

———. "Turtle Biology at Heron Island." *Australian Natural History* 15, no. 8 (1966): 262–64.

———. "Turtle Farmers of Torres Strait." *Hemisphere* 16, no. 10 (1972): 24–28.

———. "Turtle Farming and Conservation of Green Turtle (*Chelonia mydas*)." *British Herpetological Society Bulletin* 3 (1981): 36–40.

———. "Turtles and an Iguana in Fiji." *Oryx* 10 (1970): 317–22.

Bustard, H. Robert, and K. P. Tognetti. "Green Sea Turtles: A Discrete Simulation of Density-Dependent Population Regulation." *Science* 163 (1969): 939–41.

Caldwell, David K., and Archie Carr. "Status of the Sea Turtle Fishery in Florida." In *Transactions of the Twenty-Second North American Wildlife Conference (March 4–6, 1957)*, 457–63. Washington, DC: Wildlife Management Institute, 1957.

Campbell, Lisa M. "Science and Sustainable Use: Views of Marine Turtle Conservation Experts." *Ecological Applications* 12, no. 4 (2002): 1229–46.

Carr, Archie. "Alligators: Dragons in Distress." *National Geographic* 131, no. 1 (1967): 133–38.

———. "The Black Beach." *Mademoiselle*, July 1955.

———. "Caribbean Green Turtle: Imperiled Gift of the Sea." *National Geographic* 131, no. 6 (1967): 876–90.

———. "Great Reptiles, Great Enigmas." *Audubon*, March 1972, 24–34.

———. *Handbook of Turtles: The Turtles of the United States, Canada, and Baja California*. Ithaca, NY: Cornell University Press, 1952.

———. *High Jungles and Low*. Gainesville: University Press of Florida, 1953.

———. "The Navigation of the Green Turtle." *Scientific American* 212, no. 5 (1965): 79–86.

————. "The Passing of the Fleet." *American Institute of Biological Sciences Bulletin* 4 (1954): 17–19.

————. *The Reptiles.* New York: Time-Life Books, 1963.

————. *So Excellent a Fishe: A Natural History of Sea Turtles.* Garden City, NY: Natural History Press, 1967. Revised edition with author's epilogue published as *The Sea Turtle: So Excellent a Fishe*, Austin: University of Texas Press, 1984; reissued edition under the original title, with a foreword by Karen A. Bjorndal, Gainesville: University Press of Florida, 2011.

————. "Some Problems of Sea Turtle Ecology." *American Zoologist* 20, no. 3, Behavioral and Reproductive Biology of Sea Turtles (1980): 489–98.

————. "Tom Harrisson: Obituary." IUCN/SSC *Marine Turtle Newsletter* 1 (1976): 4–5.

————. "The Zoogeography and Migrations of Sea Turtles." *Yearbook of the American Philosophical Society*, 1954, 138–40.

————. *The Windward Road: Adventures of a Naturalist on Remote Caribbean Shores.* New York: Alfred A. Knopf, 1956. Reprint edition, Tallahassee: University Press of Florida, 1979.

Carr, Archie Fairly, Jr. Papers. Special and Area Studies Collections, George A. Smathers Libraries, University of Florida, Gainesville, Florida.

Carr, Archie, and David K. Caldwell. "The Ecology and Migrations of Sea Turtles, 1: Results of Field Work in Florida, 1955." *American Museum Novitates*, no. 1793 (1956): 1–23.

Carr, Archie, and Leonard Giovannoli, "The Ecology and Migration of Sea Turtles, 2: Results of Field Work in Costa Rica, 1955." *American Museum Novitates*, no. 1835 (1957): 1–32.

Carr, Archie, and Harold Hirth. "The Ecology and Migrations of Sea Turtles, 5: Comparative Features of Isolated Colonies." *American Museum Novitates*, no. 2091 (1962): 1–42.

Carr, Archie, and Robert M. Ingle, "The Green Turtle (*Chelonia mydas mydas*) in Florida." *Bulletin of Marine Science of the Gulf and Caribbean* 9, no. 3 (1959): 315–20.

Carr, Archie, and A. R. Main. "Turtle Farming Project in Northern Australia: Report on an Inquiry into Ecological Implications of a Turtle Farming Project." Department of the Special Ministry of State, Canberra, Oct. 1973.

Carr, Archie, and Larry Ogren. "The Ecology and Migration of Sea Turtles, 4: The Green Turtle in the Caribbean Sea." *Bulletin of American Museum of Natural History* 121, no. 1 (1960): 1–48.

Chaloupka, Milani, and George Balazs. "Using Bayesian State-Space Modeling to Assess the Recovery and Harvest Potential of the Hawaiian Green Sea Turtle Stock." *Ecological Modelling* 205 (2007): 93–109.

Cherfas, Jeremy. "No End to Trade in Turtles." *New Scientist* 84 (Dec. 13, 1979): 852.

———. "The Song of the Turtle." *New Scientist* 84 (Dec. 13, 1979): 880–82.

Chin, Lucas. "Obituaries and Memorial Statements: Tom Harrisson and the Green Turtles of Sarawak." *Borneo Journal* 8, no. 2 (1976): 63–69.

Christensen, R. M. "Special Report: Green Sea Turtle Farming." *Chelonia* 3, no. 2 (Aug. 1976): 4.

Collins, J. W. "Report on the Discovery and Investigation of Fishing Grounds, Made by the Fish Commission Steamer ALBATROSS during a Cruise along the Atlantic Coast and in the Gulf of Mexico: With Notes on the Gulf Fisheries." In *Report of the U.S. Commissioner of Fish and Fisheries for 1885*, app. 14. Washington, DC: Government Printing Office, 1887.

Cooley, John R. "Waves of Change: Peter Matthiessen's Caribbean." *Environmental Review* 11, no. 3 (1987): 223–30.

Culliney, John L. *Islands in a Far Sea*. Honolulu: University of Hawaii Press, 2006.

Daley, Barbara C. "Why We Still Enjoy a Plate of Turtle in Cayman Today." *Cayman Net News*, http://caymannetnews.com.ky, Aug. 17, 2007.

Daley, Ben. "Changes in the Great Barrier Reef since European Settlement: Implications for Contemporary Management." PhD diss., James Cook University, 2005.

Daley, Ben, Peter Griggs, and Helene Marsh. "Exploiting Marine Wildlife in Queensland: The Commercial Dugong and Marine Turtle Fisheries, 1847–1969." *Australian Economic History Review* 48, no. 3 (2008): 227–65.

Davidson, Osha Gray. *Fire in the Turtle House: The Green Sea Turtle and the Fate of the Ocean*. Cambridge, MA: Perseus Books, 2001.

Davis, Frederick R. *The Man Who Saved Sea Turtles: Archie Carr and the Origins of Conservation Biology*. New York: Oxford University Press, 2007.

Denevan, William M. "Bernard Q. Nietschmann, 1941–2000: Mr. Barney, Geographer and Humanist." *Geographical Review* 92, no. 1 (2002): 104–9.

Dodd, C. Kenneth, Jr. "Does Sea Turtle Aquaculture Benefit Conservation?" In Bjorndal, *Sea Turtles*, 473–80.

Donnelly, Marydele. *Sea Turtle Mariculture: A Review of Relevant Information for Conservation and Commerce*. Washington, DC: Center for Marine Conservation, 1994.

Doremus, Holly. "The Purposes, Effects, and Future of the Endangered Species Act's Best Available Science Mandate." *Environmental Law* 34 (2004): 397–450.

Doughty, Robin W. "Sea Turtles in Texas: A Forgotten Commerce." *Southwestern Historical Quarterly* 88 (1984): 43–70.

Duncan, David D. "Capturing Giant Turtles in the Caribbean." *National Geographic* 84, no. 2 (1943): 177–90.

Dutton, Peter, et al. "Composition of Hawaiian Green Turtle Foraging Aggregations: mtDNA Evidence for a Distinct Regional Population." *Endangered Species Research* 5 (2008): 37–44.

Ehrenfeld, David W. *Biological Conservation.* New York: Holt, Rinehart & Winston, 1970.

———. "Conserving the Edible Sea Turtle: Can Mariculture Help?" *American Scientist* 62 (1974): 23–31.

———. "Options and Limitations in the Conservation of Sea Turtles." In Bjorndal, *Sea Turtles*, 457–63.

Ehrenfeld, David W., and A. L. Koch. "Visual Accommodation in the Green Turtle." *Science* 155 (Feb. 17, 1967): 827–28.

Fernandez, Stephen M. "Captive-Bred Exceptions: An Unconventional Approach to Conservation under the Endangered Species Act." *University of Florida Journal of Law and Public Policy* 15 (2003): 155–93.

Florida Fish and Wildlife Research Institute, Fish and Wildlife Conservation Commission. "Green Turtle Nesting in Florida." Florida Fish and Wildlife Conservation Commission website. http://myfwc.com/research/wildlife/sea -turtles/nesting/green-turtle. Accessed June 4, 2011.

Fosdick, Peggy, and Sam Fosdick. *Last Chance Lost? Can and Should Farming Save the Green Sea Turtle? The Story of Mariculture, Ltd.—Cayman Turtle Farm.* York, PA: Irvin S. Naylor, 1994.

Gerber, Leah R., Douglas D. DeMaster, and S. P. Roberts. "Measuring Success in Conservation." *American Scientist* 88 (July–Aug. 2000): 316–24.

Gill, Alan. "Enter the Political Turtle." *ABM Review* 64, no. 3 (1974): 22–24.

Gillette, Robert. "Endangered Species: Diplomacy Tries Building an Ark." *Science* 179 (Feb. 23, 1973): 777–80.

———. "Endangered Species: Moving Toward a Cease-Fire." *Science* 179 (March 16, 1973): 1107–9.

Goode, G. Brown, ed. *The Fisheries and Fishery Industries of the United States.* 7 vols. Washington, DC: Government Printing Office, 1884–87.

———. "The Green Turtles." In *The Fisheries and Fishery Industries of the United States,* sec. 1, 150–51.

Gross, D. R. "Review of *The Turtle People,* a film by Brian Weiss." *American Anthropologist* 76 (1974): 486–87.

Hara, Leighton. "Should a Native Hawaiian Right to Take Green Sea Turtles Be Recognized under the Endangered Species Act?" Seminar paper, University of Hawaii, W. S. Richardson School of Law, Environmental Law Program, 2002.

www.hawaii.edu/elp/publications/studentarchive/s2002/hara.html. Accessed June 4, 2011.

Harrisson, Barbara. "Animal Trade, an International Issue." *International Zoo Yearbook* 14, no. 1 (1974): 13–21.

———. "International Proposals to Regulate Trade in Non-human Primates." *Primates* 13, no. 1 (March 1972): 111–14.

———. *Orang-utan*. London: Collins, 1962.

———. "Tom Harrisson: Living and Working in Borneo." *Borneo Journal* 8, no. 1 (1976): 25–30.

Harrisson, Tom. *Borneo Jungle: An Account of the Oxford Expedition to Sarawak*. London: Drummond, 1938.

———. "Breeding of the Edible Turtle." *Nature* 169, no. 4292 (1952): 198.

———. "Conservation in the Island of Borneo," in *Proceedings of the Ninth Pacific Science Congress*, vol. 7, *Conservation*, 11–12. Bangkok, Thailand: Pacific Science Association, 1958.

———. "The Edible Turtle (*Chelonia mydas*) in Borneo, 1: Breeding Season." *Sarawak Museum Journal* 5, no. 3 (1951): 593–96.

———. "The Edible Turtle (*Chelonia mydas*) in Borneo, 2: Copulation." *Sarawak Museum Journal* 6, no. 4 (1954): 126–28.

———. "The Edible Turtle (*Chelonia mydas*) in Borneo, 3: Young Turtles (in Captivity)." *Sarawak Museum Journal* 6, no. 6 (1955): 633–40.

———. "The Edible Turtle (*Chelonia mydas*) in Borneo, 4: Growing Turtles and Growing Problems." *Sarawak Museum Journal* 7, no. 7 (1956): 233–39.

———. "The Edible Turtle (*Chelonia mydas*) in Borneo, 5: Tagging Turtles (and Why)." *Sarawak Museum Journal* 7, no. 8 (1956): 504–14.

———. "The Edible Turtle (*Chelonia mydas*) in Borneo, 6: Semah Ceremonies, 1949–1958." *Sarawak Museum Journal* 8, no. 11 (1958): 482–86.

———. "The Edible Turtle (*Chelonia mydas*) in Borneo, 9: Some New Hatching Observations." *Sarawak Museum Journal* 10, nos. 17–18 (1961): 293–99.

———. "The Edible Turtle (*Chelonia mydas*) in Borneo, 11: West Borneo Numbers—The Downward Trend." *Sarawak Museum Journal* n.s. 10 (1962): 614–23.

———. *Living among Cannibals*. London: Harrop, 1943.

———. "Must the Turtle Die?" *Sunday Times* (London) *Weekly Review*, June 14, 1964, 29.

———. "Notes on Marine Turtles, 18: A Report on the Sarawak Turtle Industry (1966) with Recommendations for the Future." *Sarawak Museum Journal* 15, nos. 30–32 (1967): 424–36.

———. "The Present and Future of the Green Turtle." *Oryx* 6, no. 5 (1962): 265–69.

———. "The Sarawak Turtle Islands' 'Semah'," *Journal of the Malayan Branch of the Royal Asiatic Society* 23, no. 3 (1950): 105–26.

———. "Tagging Green Turtles, 1951–56." *Nature* 178, no. 1475 (1956): 1479.

———. "The Turtle Tragedy." *Oryx* 10, no. 2 (1969): 112–15.

Harrisson, Tom. Papers. Special and Area Studies Collections, George A. Smathers Libraries, University of Florida, Gainesville, FL.

Heimann, Judith M. *The Most Offending Soul Alive: Tom Harrisson and His Remarkable Life.* Honolulu: University of Hawaii Press, 1998.

Hendrickson, John R. "The Ecological Strategies of Sea Turtles." *American Zoologist* 20, no. 3 (1980): 597–608.

———. "The Green Sea Turtle, *Chelonia mydas* (Linn.), in Malaya and Sarawak." *Proceedings of the Zoological Society of London* 130, no. 4 (1958): 455–535.

———. "Marine Turtle Culture: An Overview." *Journal of the World Aquaculture Society* 5, nos. 1–4 (1974): 167–81.

———. "Report of Visit to Mariculture, Ltd. Turtle Farm on Grand Cayman Island, B.W.I." Report to the South Pacific Islands Fisheries Development Agency. Rome: UN Food and Agriculture Organization, 1971.

———. "Report on Hawaiian Marine Turtle Populations." In *Proceedings of the Working Meeting of Marine Turtle Specialists*, 89–95. IUCN Publications New Series Supplemental Papers, no. 20 (1969).

———. Review of *The Green Turtle and Man*, by James J. Parsons. *Science* 140 (May 24, 1963): 885.

———. "South Pacific Islands: Marine Turtle Resources." Report to the South Pacific Islands Fisheries Development Agency. Rome: UN Food and Agriculture Organization, 1972.

Hendrickson, John R., and E. Balasingham. "Nesting Beach Preferences of Malayan Sea Turtles." *Bulletin of the National Museum of Singapore* 33, no. 10 (1966): 69–76.

Hildebrand, Henry H. "A Historical Review of Sea Turtle Populations in the Western Gulf of Mexico." In Bjorndal, *Sea Turtles*, 447–53.

Hirth, Harold F. "Some Aspects of the Nesting Behavior and Reproductive Biology of Sea Turtles." *American Zoologist* 20, no. 3, Behavioral and Reproductive Biology of Sea Turtles (1980): 507–23.

———. "Synopsis of the Biological Data on the Green Turtle, *Chelonia mydas* Linnaeus 1758." FAO Fisheries Synopsis no. 58. Rome: UN Food and Agriculture Organization, 1971.

Hobart, W. L., ed. *Baird's Legacy: The History and Accomplishments of NOAA's National Marine Fisheries Service, 1871–1996.* W. L. Hobart, ed. NOAA

Technical Memorandum NMFS-F/SPO-18. Washington, DC: US Department of Commerce, June 1996.

Huff, J. Alan. "Florida (USA) Terminates 'Headstart' Program." *Marine Turtle Newsletter* 46 (1989): 1–2.

Hughes, George R. "Conservation, Utilization, Antelopes and Turtles." *Marine Turtle Newsletter* 13 (Nov. 1979): 13–14.

———. "Loggerheads and Leatherbacks in the Western Indian Ocean." *Indian Ocean Turtle Newsletter* 11 (Jan. 2010): 24.

———. "Nesting Cycles in Sea Turtles—Typical or Atypical?" In Bjorndal, *Sea Turtles*, 81–89.

Hutton, Jon M., and Nigel Leader-Williams. "Sustainable Use and Incentive-Driven Conservation: Realigning Human and Conservation Interests." *Oryx* 37, no. 2 (2003): 215–26.

Huxley, Roger. "Historical Overview of Marine Turtle Exploitation, Ascension Island, South Atlantic." *Marine Turtle Newsletter* 84 (1999): 7–9.

Ingle, Robert M. "Florida's Sea Turtle Industry in Relation to Restrictions Imposed in 1971." In *Summary of Florida Commercial Marine Landings, 1971*. Contribution no. 201, Florida Department of Natural Resources Marine Research Laboratory, St. Petersburg.

Ingle, Robert M., and Frank G. Walton Smith. *Sea Turtles and the Turtle Industry of the West Indies, Florida, and the Gulf of Mexico*. Special Publication. Miami: University of Miami, 1949.

Johnson, W. A. "Cayman Turtle Farm Ltd.: The Crock of Gold." *British Herpetological Bulletin* 2 (1980): 20–22.

Jones, Margaret. "Turtle Farming: Good Intentions but Disappointing Results." *ABM Review* 64, no. 3 (1974): 18–21.

Kaha 'ulelio, Daniel. *Ka 'Oihana Lawai'a: Hawaiian Fishing Traditions*. Honolulu: Bishop Museum Press, 2006.

Kibbe, Isaac P. "Oysters and Oyster Culture in Texas." *Bulletin of the U.S. Fish Commission* 17 (1898): 313–14.

Kinan, Irene, and Paul Dalzell. "Sea Turtles as Flagship Species: Different Perspectives Create Conflicts in the Pacific Islands." In *Marine Turtles as Flagship*, ed. Jack Frazier. Double issue of *MAST*, vol. 3, no. 2, and vol. 4, no. 1 (2005): 195–212.

King, F. Wayne. "Adventures in the Skin Trade." *Natural History* 8 (1971): 8–16.

———. "International Trade and Endangered Species." *International Zoo Yearbook* 14, no. 1 (1974): 2–13.

———. "Historical Review of the Decline of the Green Turtle and the Hawksbill." In Bjorndal, *Biology and Conservation of Sea Turtles*, 183–88.

King, F. Wayne, and P. Brazaitis. "Species Identification of Commercial Crocodilian Skins." *Zoologica* 56, no. 2 (1971): 15–70.

Kishinami, Carla. "John Roscoe Hendrickson Biography: A Daughter's Memories." Intercultural Center for the Study of Deserts and Oceans. www.cedo intercultural.org/JRHtext.htm. Accessed June 4, 2011.

Langley, Wright. "Capturing Green Turtles off Nicaragua." Master's thesis, Boston University, 1964.

———. "Green Turtle Industry Facing Extinction." *Miami Herald*, March 29, 1971.

Lewis, C. Bernard. "The Cayman Islands and Marine Turtle." In *The Herpetology of the Cayman Islands*, ed. Chapman Grant, 56–65. *Bulletin of the Institute of Jamaica: Science Series*, no. 2 (1940).

Limpus, Colin J., Emma Gyuris, and Jeffrey D. Miller. "Reassessment of the Taxonomic Status of the Sea Turtle Genus *Natator* (McCulloch, 1908), with a Redescription of the Genus and Species." *Transactions of the Royal Society of South Australia* 112 (1988): 1–9.

Limpus, Colin J., Jeffrey D. Miller, C. John Parmenter, and Duncan J. Limpus. "The Green Turtle, *Chelonia mydas*, Population of Raine Island and the Northern Great Barrier Reef, 1843–2001." *Memoirs of the Queensland Museum* 49, no. 1 (2003): 349–440.

Lipske, Michael C. "Sea Turtles Suffer as Bureaucrats Bicker." *Defenders* 52 (Aug. 1977): 228.

Lloyd, Monte, Robert F. Inger, and F. Wayne King. "On the Diversity of Reptile and Amphibian Species in a Bornean Rain Forest." *American Naturalist* 102, no. 928 (1968): 497–515.

Mack, David, Nicole Duplaix, and Susan Wells. "Sea Turtles, Animals of Divisible Parts: International Trade in Sea Turtle Parts." In Bjorndal, *Sea Turtles*, 545–62.

Matthiessen, Peter. *Far Tortuga*. New York: Vintage Books, 1975.

———. "To the Miskito Bank." *New Yorker*, Oct. 27, 1967.

McCarthy, Frederick D. "Aboriginal Turtle Hunters." *Australian Museum Magazine* 11 (March 15, 1955): 283–88.

McEvoy, Arthur F. *The Fisherman's Problem: Ecology and Law in the California Fisheries, 1850–1980*. New York: Cambridge University Press, 1986.

McNeil, Frank. "Saving the Green Turtle of the Great Barrier Reef." *Australian Museum Magazine* 11 (1955): 278–82.

Moorhouse, F. W. "Notes on the Green Turtle (*Chelonia mydas*)." In *Reports of the Great Barrier Reef Committee* 4, pt. 1, no. 1 (1931): 1–22.

Morrison, Alastair. *Fair Land Sarawak: Some Recollections of an Expatriate Official*. Ithaca, NY: Cornell University Southeast Asia Program, 1993.

Morrison, Hedda. *Sarawak*. London: MacGibbon & Kee, 1957.

Mrosovsky, Nicholas. *Conserving Sea Turtles*. London: British Herpetological Society, 1983.

———. "Dedication to Archie Carr." *American Zoologist* 20, no. 3, Behavioral and Reproductive Biology of Sea Turtles (1980): 487–88.

———. Editorial. *Marine Turtle Newsletter* 13 (Nov. 1979): 1–4.

———. "MTN 100: Looking Back, Looking Forward." *Marine Turtle Newsletter* 100 (2003): 3–4.

———. "Thermal Biology of Sea Turtles." *American Zoologist* 20, no. 3, Behavioral and Reproductive Biology of Sea Turtles (1980): 531–47.

Mrosovsky, Nicholas, Sally R. Hopkins-Murphy, and James I. Richardson. "Sex Ratio of Sea Turtles: Seasonal Changes." *Science* 225 (Aug. 17, 1984): 739–41.

Munroe, Ralph M. "The Green Turtle, and the Possibilities of Its Protection and Consequent Increase on the Florida Coast." *Bulletin of the U.S. Fish Commission* 17 (1897): 273–74.

Musgrave, Anthony, and Gilbert P. Whitley. "From Sea to Soup: An Account of the Turtles of North-West Islet." *Australian Museum Magazine* 2 (1926): 331–36.

Navid, Daniel. "Conservation and Management of Sea Turtles: A Legal Overview." In Bjorndal, *Sea Turtles*, 523–35.

Nietschmann, Bernard. *Between Land and Water: The Subsistence Ecology of the Miskito Indians, Eastern Nicaragua*. New York: Seminar Press, 1973.

———. *The Caribbean Edge: The Coming of Modern Times to Isolated People and Wildlife*. New York: Bobbs-Merrill, 1979.

———. "The Cultural Context of Sea Turtle Subsistence Hunting in the Caribbean and Problems Caused by Commercial Exploitation." In Bjorndal, *Sea Turtles*, 439–45.

———. "Hunting and Fishing Focus among the Miskito Indians, Eastern Nicaragua." *Human Ecology* 1, no. 1 (1972): 41–67.

———. *The Unknown War: The Miskito Nation, Nicaragua and the United States*. New York: Freedom House and University Press of America, 1989.

———. "When the Turtle Collapses, the World Ends." *Natural History* 83, no. 6 (1974): 34–43.

Owens, David W. "The Comparative Reproductive Physiology of Sea Turtles." *American Zoologist* 20, no. 3, Behavioral and Reproductive Biology of Sea Turtles (1980): 549–63.

———. "Introduction to the Symposium." *American Zoologist* 20, no. 3, Behavioral and Reproductive Biology of Sea Turtles (1980): 485–86.

——. "John Roscoe Hendrickson, 1921–2002." *Marine Turtle Newsletter* 99 (2003): 1–3.

Parkes, Alan Sterling. *Biologist at Large: The Background to "Off-Beat Biologist."* Published by author, 1988.

——. *Off-Beat Biologist: The Autobiography of Alan S. Parkes.* Cambridge, UK: Galton Foundation, 1985.

Parsons, James J. "English Speaking Settlement of the Western Caribbean." *Yearbook of the Association of Pacific Coast Geographers* 16 (1954): 3–16.

——. *The Green Turtle and Man.* Gainesville: University Press of Florida, 1962.

——. "The Miskito Pine Savanna of Nicaragua and Honduras." *Annals of the Association of American Geographers* 45, no. 1 (1955): 36–61.

——. "President's Address: Geography as Exploration and Discovery." *Annals of the Association of American Geographers* 67, no. 1 (1976): 1–16.

Parsons, James J., and Robert C. West. "The Topia Road: A Trans-Sierran Trail of Colonial Mexico." *Geographical Review* 31, no. 3 (1941): 406–13.

Penyapol, A. "A preliminary study of the sea turtles in the Gulf of Thailand." Paper presented at the Ninth Pacific Science Congress at Bangkok, Thailand, 1957. Published separately by the Hydrographic Department, Royal Thai Navy.

Pocock, Celmara. "Romancing the Reef: History, Heritage and the Hyper-real." PhD diss., James Cook University, 2003.

——. "Turtle Riding on the Great Barrier Reef." *Society and Animals* 14, no. 2 (2006): 129–58.

Pickett, John, and Simon Townson. "Political Problems for the Cayman Turtle Farm: Which Way Conservation?" *British Herpetological Society Bulletin*, no. 1 (1980): 18–20.

Pritchard, Peter Charles Howard. "The Conservation of Sea Turtles: Practice and Problems." *American Zoologist* 20, no. 3, Behavioral and Reproductive Biology of Sea Turtles (1980): 609–17.

——. *Tales from the Thébaïde: Reflections of a Turtleman.* Malabar, FL: Krieger Publishing, 2007.

Rebel, Thomas P. *Sea Turtles and the Turtle Industry of the West Indies, Florida, and the Gulf of Mexico.* Coral Gables: University of Miami Press, 1974.

Reichart, Henry A. "Farming and Ranching as a Strategy for Sea Turtle Conservation." In Bjorndal, *Sea Turtles,* 465–71.

Rogers, Lyman. "Conservation 70's: A Concept for Environmental Action." *Science Education* 55, no. 1 (1971): 57–60.

Rudloe, Jack. *Time of the Turtle.* New York: Penguin Books, 1979.

Rudloe, Jack, and Anne Rudloe. *Shrimp: The Endless Quest for Pink Gold.* Saddle River, NJ: FT Press, 2010.

Sand, Peter H."Commodity or Taboo? International Regulation of Trade in Endangered Species." In *Green Globe Yearbook of International Cooperation on Environment and Development, 1997,* ed. Helge Ole Bergesen and Geog Parmann, 19–36. New York: Fridtjof Nansen Institute with Oxford University Press, 1997.

Schroeder, Robert E. "Buffalo of the Sea." *Sea Frontiers* (May/June 1966): 176–83.
———. *Something Rich and Strange.* New York: Harper & Row, 1965.

Schroeder, Robert E., and W. H. Leigh. "The Life History of *Ascocotyle pachycystis* sp.n., a Trematode (Digenea: Heterophyidae) from the Raccoon in South Florida." *Journal of Parasitology* 51, no. 4 (1965): 594–99.

Schroeder, Robert E., and William A. Starck II. "Diving at Night on a Coral Reef." *National Geographic* 125, no. 1 (1964): 128–54.

Schroeder, William C. "Fisheries of Key West and the Clam Industry of Southern Florida." In *Report of the U.S. Commissioner of Fisheries for 1923,* app. 12, Washington, DC: Government Printing Office, 1924.

Schulz, Joop P. "Sea Turtles Nesting in Suriname." *Stichting Natuurbehoud Surinam (STINASU) Verhandeling* 3 (1975): 1–143.

Shelford, Robert W. C. *A Naturalist in Borneo.* London: T. Fisher Unwin, 1916.

Smith, Roger C. *The Maritime Heritage of the Cayman Islands.* Gainesville: University Press of Florida, 2000.

Smith, Virginia. "Under the Shell: Review of *Tales from the Thébaïde: Reflections of a Turtleman* by Peter C. H. Pritchard." *Times Literary Supplement,* Nov. 9, 2007.

Snover, Melissa L. "Comments on 'Using Bayesian State-Space Modeling to Assess the Recovery and Harvest Potential of the Hawaiian Green Sea Turtle Stock.'" *Ecological Modeling* 212, nos. 3–4 (2008): 545–49.

Snyder, James D. *Life and Death on the Loxahatchee.* Jupiter, FL: Pharos Books, 2002.

Spiller, Judith, and Alison Rieser. "Scientific Fact and Value in U.S. Ocean Dumping Policy." *Policy Studies Review* 6, no. 2 (1986): 389–98.

Standish, Robert I. "Meeting of the Marine Turtle Specialist Group of the Survival Service Commission, IUCN Headquarters, Morges, Switzerland, 8–10 March 1971." *Biological Conservation,* 1971, 77–78.

Stevenson, Charles H. "Preservation of Fishery Products for Food." *Bulletin of the U.S. Fish Commission* 18 (1899): 335–563.

———. "Report on the Coast Fisheries of Texas." In *Report of the U.S. Fish Commissioner for 1889 to 1891,* 373–420. Washington, DC: Government Printing Office, 1893.

———. "The United States Fish Commission." *North American Review* 176, no. 557 (1903): 593–601.

Talbot, Lee M. "Dedication to Dr. Harold J. Coolidge." *Environmentalist* 2, no. 4 (1982): 281–82.

Tisdell, Clem. "Conflicts about Living Marine Resources in Southwest Asian and Australian Waters: Turtles and Dugongs as Cases." *Marine Resource Economics* 3, no. 1 (1986): 89–109.

Udall, Stewart L. *The Quiet Crisis.* New York: Holt, Rinehart and Winston, 1963.

US Department of the Interior. "Green and Loggerhead Turtles Proposed for Foreign Endangered Species List." News Release, Jan. 4, 1974.

US Department of the Interior, Bureau of Sport Fisheries and Wildlife. "Rare and Endangered Fish and Wildlife of the United States." Sheet RAP-14, Jan. 1966.

Wallach, Bret. "In Memoriam: James J. Parsons, 1915–1997." *Annals of the Association of American Geographers* 88, no. 2 (1998): 316–28.

Webb, Graham. "Conservation and Sustainable Use: An Evolving Concept." *Pacific Conservation Biology* 8 (2002): 12–26.

Wilcox, W. A. "Commercial Fisheries of Indian River, Fla." In *Report of the U.S. Commissioner of Fish and Fisheries for the Year Ending June 30, 1896,* app. 5 Washington, DC: Government Printing Office, 1897.

Witham, Ross. "The 'Lost Year' Question in Young Sea Turtles." *American Zoologist* 20, no. 3, Behavioral and Reproductive Biology of Sea Turtles (1980): 525–30.

Witham, Ross, and A. Carr. "Returns of Tagged Pen-Reared Green Turtles." *Quarterly Journal of the Florida Academy of Sciences* 31, no. 1 (1968): 49–50.

Witham, Ross, and C. R. Futch. "Early Growth and Oceanic Survival of Pen-Reared Sea Turtles." *Herpetologica* 33 (1977): 404–9.

Witzell, Wayne N. "The Origin, Evolution, and Demise of the U.S. Sea Turtle Fisheries." *Marine Fisheries Review* 56, no. 4 (1994): 8–23.

Wood, James R., and Fern E. Wood. "Reproductive Biology of Captive Green Sea Turtles, *Chelonia mydas*." *American Zoologist* 20, no. 3, Behavioral and Reproductive Biology of Sea Turtles (1980): 499–505.

Yaffee, Steven Lewis. *Prohibitive Policy: Implementing the Federal Endangered Species Act.* Cambridge: MIT Press, 1982.

Yntema, C. L., and Nicholas Mrosovsky. "Incubation Temperature and Sex Ratio in Hatchling Loggerhead Turtles: A Preliminary Report." *Marine Turtle Newsletter* 11 (1979): 9–10.

Zangerl, R., Lupe P. Hendrickson, and John R. Hendrickson. "A Redescription of the Australian Flatback Sea Turtle, *Natator depressus*." *Bishop Museum Bulletin in Zoology* 7 (1988): 1–69.

Court Decisions

A. E. Nettleton Co. v. Diamond, 27 N.Y. 2d 182, 264 N.E.2d 118 (1970), *appeal dismissed sub nom*. Reptile Products Ass'n v. Diamond, 401 U.S. 969.

Cayman Turtle Farm, Ltd. v. Cecil D. Andrus, Secretary of the Department of Interior, 478 F.Supp. 125 (D.D.C. 1979), *affirmed per curiam*, No. 79-2031 (unpublished opinion) (U.S. Ct. Appeals, D.C. Circuit, Dec. 12, 1980).

Mariculture Ltd. v. Biggane, 369 N.Y.S. 2d 219 (N.Y.S.Ct., App. Div., 1975).

Palladio Inc. v. Diamond, 321 F. Supp. 630 (S.D.N.Y. 1970), *affirmed per curiam*, 440 F.2d 1319 (U.S. Ct. Appeals, 2d Circuit, March 26, 1971), *cert. denied*, 40 U.S.L.W. 3264 (U.S., Dec. 7, 1971).

TVA v. Hill, 437 U.S. 153 (1978).

U.S. v. Nuesca, 773 F.Supp. 1388 (D. HI. 1990).

Vermont Yankee Nuclear Power Corp. v. Natural Resources Defense Council, Inc., 435 U.S. 519 (1978).

Index

Adams, A. Maitland, 24
Adams, John Quincy, 21
African Convention on the Conservation of Nature and Natural Resources, 145
Aldabra atoll, Seychelles, 137, 141
Alert Group, IUCN, 90
Allen, Ross, 294n20
American alligator, 27, 101, 156, 161, 218
American Alligator Council, 162
American Association of Zoological Parks and Aquariums, 91
American Institute of Biological Sciences, 62, 80, 97
American Museum of Natural History, 69
American Society of Ichthyologists and Herpetologists, 58, 60, 69
American Society of Zoologists, 248
Amoroso, Emmanuel C., 200, 220, 233
Appendix I and II species under CITES, 197–98, 209–12, 214, 235, 236–39, 256
Applied Ecology Unit, Australian National University, 205
Aransas Bay, Texas, 15
Army Corps of Engineers, 173
Ascension Island, 5, 78, 99, 188
Askew, Reubin, 167, 175, 176, 177, 178, 182
Atlantic hawksbill turtle, 197
Atlantic Loggerhead Turtle Research Project, 174
Atlantic ridley turtle, 7, 131, 136, 159, 178, 184, 197
Audubon, John James, 169
Audubon Magazine, 191

Australia: and IUCN draft treaty, 196; mariculture in, 240; regulatory framework in, 184
Australian National University, 185, 205, 240
Australian Turtle Company, 31

Bahamas: green turtle population in, 21; mariculture in, 103–4, 122, 288n25
Baird, Spencer F., 14, 18
Balazs, George, 219, 238, 251, 255
Banks, Edward, 46, 47, 50, 86, 276n7
Barbour, Thomas, 279n10
Barrett, Charles, 33–34, 40
Barrier Reef Trading Company, 31
Bean, Michael, 242, 244
Bean, Tarleton H., 19
Bergen, Polly, 201–2, 203, 215
Bergen cosmetics company, 201–2
Bermuda: commercial turtling in, 112–13; green turtle population in, 21, 63
Bjorndal, Karen, 251
Bohlen, Curtis, 235
Borneo, 42, 159. *See also* Turtle Islands of Sarawak
Bramble Cay, 39
Brazaitis, Peter, 161
Brazil, turtle feeding area in, 99
breeding, captive, 6, 141, 201, 222. *See also* hatchery system
Brice, John J., 19, 20, 26
British Museum of Natural History, 49
Brongersma, Leo, 138, 144–46, 189–90, 219, 223, 226, 238
Brooke, Charles, 42, 43
Brooke, James, 42, 43

Brooke, Vyner, 42, 45, 46
Brotherhood of the Green Turtle, 8, 80–85, 279n10
Bulletin of the American Museum of Natural History, 84
Bulletin of the Raffles Museum, 93
Bureau of Commercial Fisheries, US, 211
Bureau of Sport Fisheries and Wildlife, US, 155, 284n10
Burhenne, Wolfgang, 146
Burns, John, 94
Bustard, Robert: on Australia's conservation efforts, 184, 185; and Heron Island research, 138; and IUCN, 154, 189; and Torres Islands mariculture, 143, 144, 152, 191, 205–9, 240
butchering, 24–25
Butler, William, 233, 244
Buxted Chicken Company, 105–6

Caldwell, David, 58, 69, 73, 97, 98
California, wildlife trade regulations in, 164, 215–18
California Seafood Institute, 216
Campbell, Howard ("Duke"), 186, 286n13
canning and turtle-processing factories: in Costa Rica, 183; in Florida, 24, 63; on Great Barrier Reef islands, 34, 36; in London, 78; in New Jersey, 79; in Texas, 15–17
Cape Sable beach, Florida, 102
Caribbean Conservation Corporation, 7, 81, 89, 99, 104, 133
Carnegie Museum, 80
Carr, Archie: and American alligator, 156; and Brotherhood of the Green Turtle, 8, 80; Cayman Islands research of, 62–70; and Cayman Turtle Farm, Ltd., 258; and Ebanks, 121–22; on economics of turtle fishery, 66; on egg harvesting, 145; emotional attachment to turtles, 312n35; and endangered status for green turtle, 199; and federal regulations, 238; and Florida turtle populations, 169, 170; and global

conservation strategy, 247; and Harrisson, 59–60; and Hendrickson, 84, 95, 283n21; and IUCN Red Book, 130–50, 183; and mariculture, 4, 10, 11, 96–109, 122, 134–35, 150, 200–212; and Mariculture, Ltd., 225–26; and Marine Turtle Specialist Group, 6, 95; and migration research, 99–101, 151; in *National Geographic*, 101, 105, 156; and Nietschmann, 112, 113; and Parsons, 76, 79–80, 82–83; and Sarawak tagging project, 72–75; and state regulations, 190, 219, 220; and tagging methods, 58, 60, 68–70
Carson, Rachel, 67, 85
Cayman Islands: mariculture in, 8–9; regulatory framework in, 109; research on, 62–75; tagging in, 60; turtle fishermen of, 6. *See also* Mariculture, Ltd.
Cayman Turtle Farm, Ltd.: marketing and operational plans of, 233; sale of, 263–64; and US import licenses, 4, 228, 229–46
Cayman Turtle Farm, Ltd. v. Secretaries of Interior and Commerce, 9–11, 239–46, 260–62
Cedar Keys, Florida, 24, 181
Center for Environmental Education, 247
Ceylon Journal of Science, 50
Cheatham, T. C., 169
Chelonia Institute of Truland Foundation, 250
Christian, Floyd, 179
CITES, 8, 197, 209–12, 214, 234–39, 242
Cleveland, Grover, 19
Collins, J. W., 19
Columbus, Christopher, 62
Commerce Department, US, 10–11, 211
Commission of Fish and Fisheries, 14. *See also* Fish Commission, US
Commission on Legislation, IUCN, 146
Commonwealth Development Finance Company, 224
Commonwealth Office of Aboriginal Affairs, Australia, 185

factories. *See* canning and turtle-processing factories

Hassim, Muda, 43
hatchery system: on Cayman Islands, 133; on Turtle Islands of Sarawak, 50, 81, 86, 98–99
hatch rates, 56, 201, 216, 221
Hawaii: Endangered Species Committee in, 172; turtle populations in, 93–95
Hawaiian goose, 6
Hawaiian Islands National Wildlife Refuge, 94, 186
Hawkins, John, 20–21
hawksbill turtle, 7, 118, 131, 158, 184, 196–97, 217
Hayek, F. A., 106
Hendrickson, John R.: and captive breeding, 141; and Carr, 84, 283n21; and global conservation strategy, 79, 135–36, 259–60; Hawaii research by, 93–95, 186; and IUCN, 189; and mariculture, 96, 98–99, 143, 152, 187, 200, 248–49, 255; and Maricul-ture, Ltd., 190–91; and Marine Turtle Specialist Group, 95; and Monel tags, 69, 72–75; and Parsons, 77; review of Parsons' book by, 83–84; and Sarawak tagging project, 72–75; and state regulations, 203–4, 215; and Turtle Islands of Sarawak research, 11, 53–61, 74
Heron Island, 31, 34, 36, 39, 49, 52
Herrera, Andrés, 131
High Jungles and Low (Carr), 76, 81, 112
Hildebrand, Henry, 131
Hirth, Harold: and Aldabra atoll, 141; career of, 286n13; Carr's influence on, 259; and South Pacific Fisheries Program, 94, 132, 152; Tortuguero research by, 69, 81, 135
Hodges, Randolph, 175, 179–80
Holloway, Colin, 136, 137, 144, 145–47, 153, 189
Honegger, René, 153, 154
Hopkins-Murphy, Sally, 249
Hornaday, William, 160
House of Refuge Museum, 171
Hughes, George, 219, 222, 252
Huxley, Julian, 48, 88, 89

imports: Florida regulations on, 9, 179, 183, 192, 298n30; prohibition on, 4; US licenses for, 4, 228, 229–46; US regulations for, 4, 7–9, 10, 28–29. *See also* exports
Incer, Jaime, 253
incubation period, 53
Indian River Lagoon, 21–22, 24, 27
Inger, Robert F., 159
Ingle, Bob, 167–70, 173, 175, 177, 179–81
Institute for Medical Research, Kuala Lumpur, 54
Institute of Economic Affairs, 106
Inter-American Press Association, 81
Interior Department, US: and CITES, 234; and endangered classification, 154–59, 171–73, 265; Endangered Species Act responsibilities of, 211; and IUCN draft treaty, 195, 197; and mariculture, 10–11
International Union for Conservation of Nature (IUCN): and definition of endangered species, 194; founding of, 276n18; and Mariculture, Ltd., 218–24, 225–26; mariculture principles of, 98, 223–24, 266–67; Nairobi meeting (1963), 89–91; Red Data Book, 2, 5, 90, 130–50, 184; and state regulations, 215–18; status categories of, 153. *See also* Marine Turtle Specialist Group
Izaak Walton League, 173

Jamaica, green turtle population in, 63
Japan and IUCN draft treaty, 196
Johnson, Johnny, 220, 225–26, 257
Jones, Carleton, 217–18, 224–26, 233, 238, 239–46
Journal of Zoology, 220
judicial case history: endangered species designations, 310n39; mariculture ban, 9–11, 239–46, 260–62; New York wildlife trade regulations, 163
Jupiter Island, Florida, 174, 176, 178

Kemp's ridley turtle, 7, 159, 197
Kenya, turtle exports from, 79, 145

Parsons, James J.: and Carr, 76, 79–80, 82–83; and Nietschmann, 110–11; research by, 11, 76–85; and Tortuguero research station, 100–101
Pearke, Charles, 24
Pepper, William M., Jr., 80–81
personal philosophy influencing scientific judgment, 3–4, 12
poachers, 103, 152
Ponce de León, Juan, 21
populations: in Bahamas, 21; in Bermuda, 21, 63; in Florida, 169, 170; of green turtles, 63; in Sarawak, 48–53, 56–57, 86–87
Powers, Joshua B., 81
Pratt, John H., 9, 10, 244, 260–62
Principles and Recommendations on Commercial Exploitation of Sea Turtles, IUCN, 223–24, 233, 266–67
Pritchard, Peter C. H.: appointment of, 139, 144, 189–90; on Carr, 312n35; and federal regulations, 238; and global conservation strategy, 250, 256, 259; and IUCN mariculture principles, 219; and IUCN Red Data Book, 152
Proceedings of the Zoological Society of London, 50, 77
Prohibitive Policy: Implementing the Federal Endangered Species Act (Yaffee), 2, 10
prohibitory regulations, 2–5
Providencia Island, 110
public opinion, 91–92. *See also* cultural attitudes
Puerto Cabezas, Nicaragua, 109, 127, 129, 139
Purchon, Richard D., 53, 59

Queensland Department of Harbor and Marine, 40
Queensland Parliament, 36, 38, 39
Queensland Society for the Prevention of Cruelty to Animals, 40

ranching schemes, 244
rare classification, 153
Reed, Nathaniel, 195, 196, 197

regulatory framework: for alligator harvesting, 156; in Australia, 184; in California, 164, 215–18; on Cayman Islands, 109; in Costa Rica, 65–66, 81, 100; under Endangered Species Act, 213–28; in Florida, 9, 97–98, 156, 165–82, 298n30; on Great Barrier Reef islands, 35, 38–39; and IUCN, 146, 155; in Massachusetts, 164; in New York, 162, 163, 214; in Nicaragua, 108–9, 123, 168; prohibitory regulations, 2–5; salvage law, 20–21; state, 190, 219, 220; on Turtle Islands of Sarawak, 46
Reichart, Henk, 223, 242, 243, 247, 254
Reptile Products Association, 164
Reptiles, The (Carr), 8, 80, 96, 154, 186
restocking strategies, 99, 171, 296n16
Revenue Service, US, 21
Richardson, James, 249
Roosevelt, Theodore, 160
Royal Asiatic Society Journal, Malayan Branch, 50
Royal Society for the Prevention of Cruelty to Animals, 37
Rudloe, Jack, 178

Sahrhage, D. W., 140, 141
salvage law, 20–21
San Andres Island, 110–11
sandhill cranes, 3, 154
Sarawak. *See* Turtle Islands of Sarawak
Sarawak Museum, 11, 50
Sarawak Museum Journal, 58, 87
Satang Besar, 42, 45. *See also* Turtle Islands of Sarawak
Sauer, Carl, 282n12
Savage Civilization (Harrisson), 48
Schoning, Robert, 237
Schroeder, Jean, 101
Schroeder, Robert, 7–9, 101–9, 122–23, 140, 144, 190–91
Schroeder, William C., 24, 26
Schulz, Joop: and Brongersma, 145; and global conservation strategy, 254, 256; and IUCN mariculture principles, 219, 223; and Mariculture,

Ltd., 215, 226; and Suriname mariculture, 188, 215, 242
Scott, Peter, 6, 12, 88, 89, 90, 136, 149
Scott, Robert F., 6, 88
Sea Farms, Inc., 165
Sea Life Park, Hawaii, 186
Semah celebrations, 43, 50, 79, 92
Sequeira, Manuel, 108–9
Seychelles, mariculture in, 137
Shackleton, Eddie, 47
Shelford, Robert, 275n3
shoe manufacturers, 9, 85, 88, 160–61
shrimp fishermen, 11, 173, 297n23
Sierra Club, 215
Silent Spring (Carson), 67, 85
Simon, Marlin, 141
Smith, Frank G. Walton, 168, 176, 277n20
Smith, Hugh M., 20, 27, 28
snail darter, 3, 310n39
So Excellent a Fishe (Carr), 112, 135, 136, 143, 186, 258, 282n15
soup. *See* turtle soup
Southeastern Fisheries Association, 171, 173, 176
Southern Steamship Company, 16
South Pacific Commission, 99
South Pacific Fisheries Program, 93
sponges, 19, 20
Squier, Ephraim George, 111
State Department, US, 177
Stearns, Silas, 14
Stebbins, Robert C., 53, 58, 73
Stevens, Robert, 236
Stevenson, Charles H., 16, 17
St. Kilda, 48
Sunday Times of London, 92
Supreme Court, US, 164, 241, 310n39
Suriname, mariculture in, 188, 242, 254
Suriname Forest Service, 215
Survival Service Committee, IUCN, 89, 98, 132–36, 216
sustainable-use strategies, 4
Sweat, Don, 165–67

tagging: and global conservation strategy, 252; on Great Barrier Reef islands, 34–35, 49, 54; in Hawaii, 94; rewards for return of tags, 69,

115; in Tortuguero research station, 100; on Turtle Islands of Sarawak, 11, 54–61
tags, Monel (cow-ear), 54–55, 60, 69
takes, incidental, 238
Talang Talang Besar, 42, 84. *See also* Turtle Islands of Sarawak
Talang Talang Kechil, 42. *See also* Turtle Islands of Sarawak
Talbot, Lee, 197
Tasbapauni, Nicaragua, 124–25, 150, 168
Taylor, Zachary, 111
Texas: commercial fisheries in, 14–19; decline in green turtle catches in, 17; fisheries legislation in, 18
Texas Fish and Oyster Commission, 18
Thomas, Jerry, 175
Thompson, Norberg, 24, 124, 165
Thompson Enterprises, 71
Tisdell, Clem, 264
Torres Strait Islands: breeding season on, 36; mariculture on, 184–85, 191, 205–9, 240, 263–64; nesting beaches on, 39, 143
Tortuguero research station, Costa Rica: and CITES, 240, 253; hatchery system at, 81, 133; Nietschmann at, 112–23; scientific research at, 62–70; tagging research at, 100
tourism, 34, 39, 146, 167, 264
tripartite agreement, 139, 148–50, 183
Tromelin, rookery beaches on, 5
Truland Foundation, 250
Turtle Board of Management, Sarawak, 46, 57, 86, 264
Turtle Bogue, 63, 64, 279n6
turtle farming. *See* mariculture
Turtle Islands of Sarawak, 42–61; and Banks, 44–46; Harrisson's work on, 46–61; hatchery system on, 98–99; Hendrickson's work on, 53–61; nesting population of, 48–53, 56–57; scientific research on, 48–61
Turtle People, The (film), 253
turtle riding, 34, 39
turtle soup: culture and attitudes regarding, 79–80; demand for, 5–6, 63, 78, 85, 103; geography of, 76–85;

turtle soup *(cont.)*
popularity of, 24, 29, 77; and US
mariculture import prohibition, 4
turtle turners. See *veladores*
TVA v. Hill, 310n39

Udall, Stewart, 154, 155
Ulrich, Glenn, 259
Union Creek Experimental Green
Turtle Culture Project, 288n25
United Nations Conference on the
Human Environment (1972), 195
United Nations Food and Agriculture
Organization, 93–94, 109, 135, 138,
140, 142, 152
University of Arizona, 186
University of California at Berkeley, 53,
111
University of Hawaii at Manoa, 83, 93
University of Malaya, 53
University of Western Australia, 206
US Agency for International Develop-
ment, 140

veladores, 66, 68
Vesely-Forte, Inc., 196

*Waikna: Adventures on the Miskito
Shore* (Squier), 111
Wallace, Alfred Russel, 42
Wallace, H. E., 172
Weiss, Brian, 253
West, Robert, 110
White House Council on Environmen-
tal Quality, 195
White Rajahs, 42
Whitley, Gilbert, 31, 33, 39
Wild and Scenic Rivers Act (1968), US,
297n25
Windward Road, The (Carr), 8, 72, 80,
97, 131, 279n10, 282n15
Witham, Ross, 169, 174, 248, 249, 259
Wood, Fern, 248, 256, 259
Wood, Jim, 190, 200, 225–26, 248,
256, 259
Woods Hole, Massachusetts, 18
World Conference on Sea Turtle
Conservation, 251
World Wildlife Fund, 88, 89, 90, 183,
191, 247, 258

Yaffee, Steven L., 2–5, 10
Yglesias, Guillermo, 148–49